GW01260254

Technology and the Culture Progress in Meiji Japan

In this book David Wittner situates Japan's Meiji Era experience of technology transfer and industrial modernization within the realm of culture, politics, and symbolism, examining how nineteenth-century beliefs in civilization and enlightenment influenced the process of technological choice.

Through case studies of the iron and silk industries, Wittner argues that the Meiji government's guiding principle was not simply economic development or providing a technical model for private industry as is commonly claimed. Choice of technique was based on the ability of a technological artifact to import Western 'civilization' to Japan: Meiji officials' technological choices were firmly situated within perceptions of authority, modernity, and their varying political agendas. Technological artifacts could also be used as instruments of political legitimization. By the late Meiji Era, the former icons of Western civilization had been transformed into the symbols of Japanese industrial and military might.

A fresh and engaging re-examination of Japanese industrialization within the larger framework of the Meiji Era, this book will appeal to scholars and students of science, technology, and society as well as Japanese history and culture.

David G. Wittner is Associate Professor of Asian History at Utica College in Utica, New York.

Routledge/Asian Studies Association of Australia (ASAA) East Asia Series

Edited by Tessa Morris-Suzuki and Morris Low

Editorial Board: Professor Gemerie Barmé (Australian National University), Professor Colin Mackerras (Griffith University), Professor Vera Mackie (University of Melbourne) and Associate Professor Sonia Ryang (Johns Hopkins University).

This series represents a showcase for the latest cutting-edge research in the field of East Asian studies, from both established scholars and rising academics. It will include studies from every part of the East Asian region (including China, Japan, North and South Korea and Taiwan) as well as comparative studies dealing with more than one country. Topics covered may be contemporary or historical, and relate to any of the humanities or social sciences. The series is an invaluable source of information and challenging perspectives for advanced students and researchers alike.

Routledge is pleased to invite proposals for new books in the series. In the first instance, any interested authors should contact:

Professor Tessa Morris-Suzuki
Division of Pacific and Asian History
Research School of Pacific and Asian Studies
Australian National University
Canberra, ACT0200 Australia

Professor Morris Low
Department of the History of Science and Technology
Johns Hopkins University
3505 N. Charles Street
Baltimore, MD 21218, USA

Routledge/Asian Studies Association of Australia (ASAA) East Asia Series

1   **Gender in Japan**
    Power and public policy
    Vera Mackie

2   **The Chaebol and Labour in Korea**
    The development of management strategy in Hyundai
    Seung Ho Kwon and Michael O'Donnell

3 **Rethinking Identity in Modern Japan**
Nationalism as aesthetics
Yumiko Iida

4 **The Manchurian Crisis and Japanese Society, 1931–33**
Sandra Wilson

5 **Korea's Development Under Park Chung Hee**
Rapid industrialization, 1961–1979
Hyung-A Kim

6 **Japan and National Anthropology**
A critique
Sonia Ryang

7 **Homoerotic Sensibilities in Late Imperial China**
Wu Cuncun

8 **Postmodern, Feminist and Postcolonial Currents in Contemporary Japanese culture**
A reading of Murakami Haruki, Yoshimoto Banana, Yoshimoto Takaaki and Karatani Kōjin
Murakami Fuminobu

9 **Japan on Display**
Photography and the emperor
Morris Low

10 **Technology and the Culture of Progress in Meiji Japan**
David G. Wittner

Technology and the Culture of Progress in Meiji Japan

David G. Wittner

To Chris,
Once my support network,
always my support network!

Routledge
Taylor & Francis Group

LONDON AND NEW YORK

First published 2008
by Routledge
2 Park Square, Milton Park, Abingdon, Oxon, OX14 4RN

Simultaneously published in the USA and Canada
by Routledge
270 Madison Ave, New York NY 10016

Routledge is an imprint of the Taylor & Francis Group, an informa business

Transferred to Digital Printing 2009

Typeset in New Times Roman by
Keystroke, 28 High Street, Tettenhall, Wolverhampton

British Library Cataloguing in Publication Data
A catalogue record for this book is available from the British Library

Library of Congress Cataloging in Publication Data
Wittner, David G.

Technology and the culture of progress in Meiji Japan / David G. Wittner
p. cm. – (Routledge/Asian Studies Association of Australia (ASAA)
East Asia series ; 10)
Includes bibliographical references and index.
1. Industrialization–Japan. 2. Technology–Social aspects–Japan.
3. Industries–Social aspects–Japan. 4. Japan–Economic conditions–1868–1918.
I. Title
HC462.7.W57 2007
338.095209′034–dc22
2007016235

ISBN10: 0–415–43375–4 (hbk)
ISBN10: 0–415–56061–6 (pbk)
ISBN10: 0–203–94637–5 (ebk)

ISBN13: 978–0–415–43375–4 (hbk)
ISBN13: 978–0–415–56061–0 (pbk)
ISBN13: 978–0–203–94637–4 (ebk)

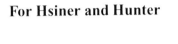

For Hsiner and Hunter

Contents

List of illustrations   xi
Acknowledgments   xiii
List of abbreviations   xv
Glossary   xvii

1   Introduction: Meiji modernization revisited   1

2   Tradition and modernization   19

3   Iron machines and brick buildings: the material culture
    of silk reeling   43

4   Smelting for civilization: technological choice and the
    modernization of the iron industry   72

5   *Bunmei kaika to gijutsu*: technology's role in
    'civilization and enlightenment'   99

6   Conclusion: *from technological determinism to
    techno-imperialism*   125

    *Appendix I: Proposals for the location of Kamaishi
       Ironworks*   130
    *Appendix II: Distribution and productivity of filatures
       by prefecture*   132
    Notes   136
    Bibliography   174
    Index   195

Illustrations

Figures

2.1	*Zaguri*	20
2.2	Woman reeling silk with belt-driven *zaguri*	21
2.3	Rear view of belt-driven *zaguri*	22
2.4	Re-reeling machine	23
2.5	Foot-powered traditional reeling machine with *tavelle*	40
2.6	Water-powered traditional reeling machines with *tavelle*	41
3.1	Raw silk packaging label: Hecht, Lilienthal & Co.	45
3.2	French-style cast iron reeling frame from Meiji silk reeler's manual	56
3.3	Western-style factory from Meiji silk reeler's manual	57
3.4	Examples of filature exterior materials	58
3.5	*Chambon* method of *croisure*, Tachi Saburō	60
3.6	*Chambon* method of *croisure*, Itō Moemon	61
3.7	*Tavelle* method of *croisure*, Tachi Saburō	62
3.8	*Tavelle* method of *croisure*, Itō Moemon	63
5.1	Raw silk packaging label, 'Ammunition Brand,' Katakura & Co. Filatures.	113
5.2	Raw silk packaging label, 'Propellor Brand,' Katakura & Co. Ltd.	114

Tables

3.1	Raw silk exports	46
3.2	Commendations received at 1873 International Exhibition	70
4.1	Location of blast furnaces in Kamaishi area	75
5.1	Iron and steel imports, 1874–1900	117
5.2	Disposal of government enterprises	121

Maps

0.1	Map of Japan	xxi
4.1	Map of Kamaishi area showing blast furnace location	76

Acknowledgments

This project began while I was at The Ohio State University where I benefitted from the advice, intellectual support, and friendship of James R. Bartholomew, Philip C. Brown, Christopher A. Reed, and Maureen Donovan. Each has contributed to the overall development of this project. Since that time, I have continued to receive valued counsel and advice from these fine scholars and friends. I am also grateful to Morris F. Low (University of Queensland) for his patience and wisdom as I worked through my ideas and draft manuscripts.

In some ways I feel like one of the Meiji officials whom I discuss in this book. Based on a series of introductions by my good friend Okamoto Kōichi (Waseda University) and Simon Partner (Duke University), I was welcomed at Tokyo University's Institute for Social Science Research (*Shaken*). There, I received the support of Kudō Akira, and thank him for inviting me to his *kenkyūshitsu* and his home. His advice while at *Shaken* greatly contributed to my time in Japan. Similarly, I thank Sashinami Akiko and Suzuki Jun (Tokyo University) for their counsel and guidance. My time at *Shaken* was facilitated by many dedicated staff members and the institute's librarians, to all of whom I owe a debt of gratitude. I would also like to thank the library staff at Tokyo University Faculties of Economics and Law, and Fujii Sachiko of the Meiji Shinbun Zasshi Bunko for their patience and assistance on my original and subsequent research trips. Most recently, Shimizu Kenichi of the National Science Museum in Tokyo who made my visit to Tomioka Filature possible.

Since my first research trip in 1996, Murahashi Katsuko of the Keidanren has been a valued colleague. Koide Izumi, International House of Japan, and Harada Noriko, National Science Museum, Ueno, Tokyo, were instrumental in helping me locate materials held in various museum collections. I greatly benefitted from the generosity and knowledge of curatorial staff at the Gunma Prefecture Museum of Technology in Takasaki and the Kamaishi Iron and Steel History Museum in Kamaishi City. The members of the Japan Association of Mining and Metallurgy Research (*Kinzoku Kōzan Kenkyūkai*) and especially Murata Sunao, Hatakeyama Jirō, and Onozaki Satoshi made my research regarding Kamaishi Ironworks quite satisfying. Throughout my stay in Japan, Jonathan Lewis (now of Hitotsubashi University) and his wife Maho made my trip all the more enjoyable. I greatly appreciate their friendship and wish to thank them for all their help.

Some people say that good friends need not be thanked because they know they are appreciated. I must disagree. Christopher Gerteis (Creighton University), Jason Karlin (Tokyo University), and Sabine Frühstück (University of California, Santa Barbara) formed my core support network. We shared camaraderie and many cups of coffee, tea, and sake discussing each other's work or life in general. Each has contributed intellectually and emotionally to this project in some way. Parts of the original manuscript were written while in Jim Bartholomew's graduate seminar where I benefitted from the insightfulness of Sumiko Otsubo Sitcawich (Metropolitan State University) and Christienne Hinz (Southern Illinois University, Edwardsville).

The original project was generously supported by the Fulbright Foundation and the Japan–United States Education Commission. I would like to thank the dedicated staff at JUSEC, and especially Samuel Shephard, executive director, and Toyama Keiko, who handled the details of my stay in Japan. The Japan–America Society of Chicago and The Ohio State University's Graduate Summer Research Fellowship program also generously contributed to my early research. The final stages of this project have been supported by the Faculty Leadership Fund and Crisafulli Fund at Utica College.

Last and certainly not least, I would like to thank those to whom I owe my greatest debt, my family. My parents, Murray and Phyllis Wittner, have been a constant source of support. Most of all, I thank my wife Hsin Hsin who has been by my side and endured this project since its inception; her support has never waned. Without her encouragement and patience, the experience of writing this book would have been significantly more difficult and far less enjoyable. And to Hunter, who spent his first two years with this project, I say thank you for growing with me.

Abbreviations

F.O.	*Foreign Office Papers*
FSK	*Kamaishi seitetsujo shichijū nenshi* (Fuji Seitetsu Kabushikigaisha Kamaishi Seitetsujo)
FYZ	*Fukuzawa Yukichi zenshū*
IHKM	*Itō Hirobumi kankei monjo*
KH	*Kyōshinkai hōkoku, kenshi no bu*
KEH	*Kōbusho enkaku hōkoku*
KNGS	*Kōza Nihon gijutsu no shakaishi*
MPICE	*Minutes and Proceedings of the Institution of Civil Engineers*
MZZKSS	*Meiji zenki zaisei keizai shiryō shūsei*
OHHK	*Ōkoku hakurankai hōkokusho: kōgyō denpa hōkokusho*
OHHS	*Ōkoku hakurankai hōkokusho: sangyōbu*
OSKM	*Ōkuma Shigenobu kankei monjo*
OTK	*Ōshima Takatō kōjitsu*
RSJCI	*Kamaishi oyobi Sennin tetsuzan junshi hōkoku* (Rinji Seitetsu Jigyō Chōsa Iinkai)
SEDS	*Shibusawa Eiichi denki shiryō*
SID	*Segai Inouekō den*
SYZ	*Sugiura Yuzuru Zenshū*
TSS	*Tomioka seishijoshi*
TTH	*Tetsu to tomo ni hyakunen*

Japanese Names

All names are given according to Japanese convention, i.e., surname first, given name second, except in instances where author preference dictates otherwise.

Glossary

akome	red iron sand
bakufu	literally 'tent government;' shogunate
bakuhan	literally bakufu and domain. Describes the Edo period (1603–1868) goverment
bunmei kaika	civilization and enlightenment (movement)
bunmei no riki	conveniences of civilization
daibunmei *daikaika no kuni*	great civilized and enlightened countries, used to describe France and Great Britain
daijō	senior deputy minister
daimyō	ruler of a domain; feudal lord
dajōkan	Council of State
doguri	early silk reeling device
fukoku kyōhei	'Rich Nation, Strong Army'; a Meiji-era government slogan used to promote industrial development and nation building
fukokuron	rich nation thesis or thesis on national prosperity
fumifugio	foot-operated bellows
gaikoku bugyō	Foreign Magistrate
Gaimushō	Ministry of Foreign Affairs
garabō	early cotton spinning machine known as the 'rattling spindle' because of its rather noisy operation
geta	Japanese clogs
gonnokami	deputy chief
habutae	a plain silk fabric
han	semi-autonomous domain(s)
hanbatsu	clan-faction (politics). Term used to describe factionalism and the political dominance of men from Satsuma and Chōshū in Meiji and early Taishō era government.
hata	a machine or mechanical device
hodoana	the hole through which the furnace master can inspect conditions within a tatara furnace
ikensho	written opinion piece

inoshishi	literally 'wild boar,' used to describe a person who acts quickly without considering the consequences
jinrikisha	rickshaw
jōi	usually *sonno jōi*, literally 'Revere the Emperor, Expel the Barbarian,' an anti-foreign slogan and movement of the late Edo era
jōkamachi	castle towns that served as regional, i.e., domain, capitals
kago	seat mounted on two poles carried by porters
kaibutsu	monster
kaigun	navy
Kaisei Gakkō	School of Western Studies
Kangyō kyoku	Industrial Promotion Bureau
kanji	Chinese characters
kanna nagashi	method of mining that uses rushing water to extract iron sand from soft granite
kera	a lump of steel and iron left at the end of the smelting process
kikai	machine
Kōbu daigakkō	College of Engineering
Kōbushō	Ministry of Public Works
Kōbushōdayū	Vice Minister of Public Works
Kōbushō kankōryō	Office for the Promotion of Industry
Kōbushō kōzankyoku	Ministry of Public Works Mining Bureau
Kōgakuryo	Engineering Bureau
Kōgyōharaisage	the sale of government industry
kokutai	national polity
Kōzan kyoku	Mining Bureau
kunyomi	Japanese pronunciation or reading of a Chinese character or *kanji*
masa kogane	black iron sand
Minbushō	Ministry of Civil Affairs
minkan	the people
mizunuki	drain or lower-level of a mine
Monbushō	Ministry of Education
murage	ironmaster
Naimushō	Ministry of Home Affairs
nodatara	early iron furnace situated in a field
noile	waste silk thread
noshi	coarse thread covering the outer layers of a silkworm cocoon
Nōshōmushō	Ministry of Agriculture and Commerce
Ōkurashō	Ministry of Finance

onyomi	Chinese pronunciation or reading of a Chinese character or *kanji*
rangaku	Dutch Learning
Rokumeikan	building designed by architect Josiah Conder that came to symbolize the Westernization of Japan
sangi	Grand Councilors
Sei'in	Central Chamber of the government
sentetsu	pig iron
shamisen	a three-string musical instrument resembling a banjo
shinpo	progressing or advancing in civilization
shokusan kōgyō	policy to increase production and promote industry, program of industrialization
sozei gonnokami	Commissioner of Revenue
takadano	building in which a tatara furnace was housed
tatara	traditional Japanese furnace for smelting iron which resembles a trough
tebiki	wooden-framed, hand-cranked reeling device
tenbindatara	tatara furnace blown by a balance bellows
tenbinfuigo	balance bellows
yatoi	foreign employee, adviser
zaguri	traditional silk reeling frame
zaibatsu	conglomerate corporation

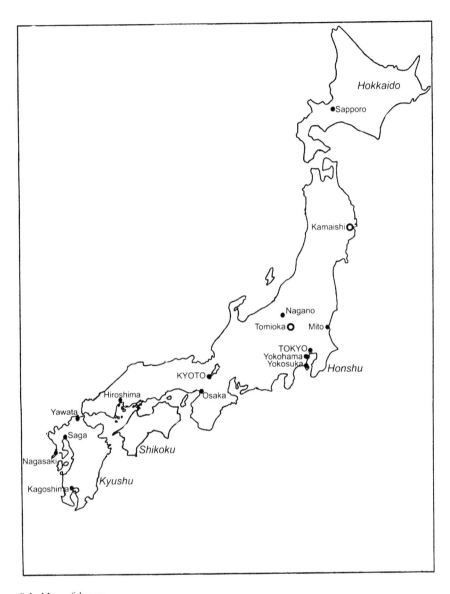

0.1 *Map of Japan*

1 Introduction

Meiji modernization revisited

In the decades following the Meiji Restoration of 1868, the fabric of Japanese society changed. Political systems, economics, industry, transportation, dress, social and cultural values; in short, nearly every aspect of life in Japan was affected. Key to these changes was the new Meiji government's attitude toward the adoption of Western technologies and the values which these objects and ideas from the 'civilized' countries of Europe would transmit to Japan. The degree of interconnectivity between the adoption of Western technologies, the Japanese enlightenment movement,[1] and the government's need for political validation makes it nearly impossible to study any one topic without at least recognizing this relationship. Traditionally, the study of technology transfer to the iron and silk industries has been looked at either in isolation or as case studies of developmental economics.[2] Rarely has Japanese industrialization been examined from the perspective of political validation, never from the perspective of ideological importation.[3]

This book is an examination of the interaction between technology and ideology in nineteenth-century Japan. It illustrates how choice of technique in the iron and silk industries – the selection of methods by which the Meiji government sought to mechanize these once traditional industries – was guided by individual beliefs in 'modernity,' 'civilization,' and 'enlightenment.' It situates government-sponsored efforts at industrialization within the discourse of *bunmei kaika* ideology, the so-called movement to attain 'civilization and enlightenment.' My aim is to demonstrate that the pre-eminent concerns for choice of technique and technology transfer in the iron and silk industries were more related to the government's need for self-validation vis-à-vis the soon-to-be-abolished domains (*han*), former shogunate (*bakufu*), and foreign treaty powers than simply being related to economic development or building a '*Rich Nation* and *Strong Army*' in strict politico-economic terms. Moreover, the government's sale of its 'model factories' to private firms (*haraisage*) was situated within the Meiji leaders' acceptance of contemporary beliefs in an historical development of civilizations; sales were not based strictly on economic rationalism. As Japan became an international political player in the 1880s, material manifestations of Western 'civilization' were supplanted by the next ideological stage, imperialism.

The story of how the Japanese government 'modernized' the country through the adoption, and later adaptation, of Western technologies is well known, or so we believe. In an effort to quickly revise the unequal treaties, Meiji leaders set out on a tour of Europe and the United States, the Iwakura Mission, 1871–73, where they learned about the West. They hired foreign advisers to speed along the modernization process, and sent students abroad who eventually replaced the foreign tutors. By the 1890s this was largely accomplished. Japan had a new form of government, a rapidly growing industrial economy, a substantial transportation and communications infrastructure, a new system of public education, a military establishment, and diplomatic relations with the major world powers which led to the revision of the unequal treaties. This transformation, however, was not based on a well-developed plan. It was the product of numerous ad hoc solutions to the myriad problems that confronted the Meiji leaders on a seemingly daily basis. As a result, many solutions were multi-functional, providing for any number of situations. In the case of the government's attempts to mechanize and modernize the silk reeling and iron mining and smelting industries, no statement could be more accurate.

Why iron and silk? These two naturally occurring raw materials are as different in their refined states as they are from each other. The processes by which iron and silk are transformed into value-added commodities, the traditional laborers in each industry, their markets, and the social status representations of these materials, just to mention a few aspects, are completely dissimilar. Yet the two are often studied as examples of unsuccessful technology transfer by economic historians and developmental theorists who wish to use Japan's nineteenth-century industrialization as a model for newly industrializing economies.[4]

Failure and success, however, are not valid, or are at least highly subjective measures of a technology's transferability when one does not recognize the historical actors' motivations in selecting a technology. Like many aspects of engineering in general, choice of technique and technology transfer can be considered personal processes.[5] Recognizing individual motivation, or appreciating that choice of technique may be guided by one person or a small group of people with varying agendas – agendas that may or may not be the same as their publicly stated goals – is essential for understanding the mechanisms by which technologies are selected and transferred. I focus on iron and silk, not as comparative case studies of failed technology transfer as is common practice, but specifically because in many respects, iron and silk are as different as the circumstances surrounding the Meiji government's attempts to modernize the respective industries.

From an industrial perspective, iron and silk represent heavy and light industry, and are revealing of the different attitudes that the government had toward each sector. Throughout Japanese history, iron smelting and blacksmithery were the crafts of a select group of artisans. While iron smelting was often done during the agricultural off-season, it was not a popular by-employment. In fact, the technologies were not widely disseminated as many artisans guarded their trade secrets. Sericulture and silk reeling, however, were traditional by-employments of the agricultural sector that, until the end of the nineteenth century, competed for labor during the peak of the rice growing season.[6] Related technologies were widely disseminated and implemented. As such, government officials sought to promote the modernization of these industries on a popular level. Raising the quality (and quantity) of raw silk was considered an act for the good of the people and for the good of the nation. The ubiquitous *zaguri* silk reeling frame could be found in many farmers' homes; the same can not be said for the ironmaster's *tatara* furnace.

For their part, officials within the Meiji government recognized this basic difference and, although some bureaucrats would overestimate the ease with which industrial iron smelting technology could be transferred, the government's position is well articulated. In his report on the 1873 International Exhibition in Vienna, Ministry of Civil Affairs (*Minbushō*), official Sano Tsunetami stated that, unlike the silk industry where improvements are necessarily done on a small-scale basis by private producers, improvements to mining and metallurgy are 'simple' because it is large scale and only involved the importation of large machinery by the government.[7] Introduced for the good of the people and the nation, improvements to mining and metallurgy were not, however, to be performed *by the people*.

Following this reasoning, the improvement of silk reeling was promoted on the popular level. Government directives and memorials targeted the people. First the merchants were urged to abandon their greedy and deceitful trade practices; later the producers were told to improve their reeling methods for the good of the nation.[8] When the government, in 1870, decided to open its own filature in the village of Tomioka in Gunma prefecture, the professed function was for it to serve as a national model of modern industry. Tomioka's stated purpose was to disseminate the latest Western-style silk reeling technologies to the people. Private producers and merchants would come to the factory to learn and study the new technologies; women reelers from around the country would come to Tomioka to be trained in the latest Western methods.[9]

When, in 1874, bureaucrats and ministers in the Ministry of Public Works (*Kōbushō*) decided to nationalize the Kamaishi mines in Iwate prefecture and erect a Western-style ironworks on the site, there was no prior call to the people to improve their methods of iron smelting.[10] There was no invitation to the populace to later visit the facility to learn the latest Western mining and iron smelting techniques. Although a public venture, Kamaishi Ironworks was not, per se, open to the public. None of the popular recognition and fanfare that enveloped nearly every aspect of Tomioka Silk Filature can be found with regard to any aspect of Kamaishi Ironworks.

Yet Kamaishi and its associated mines were also being developed for the 'good of the nation.' In January 1872, Vice-Minister of Public Works, Yamao Yōzō, sent a petition to the Central Chamber (*Sei'in*) designed to rouse the government to action. He stated that mining was the basis of a 'rich imperial nation.' In an effort to awaken the spirit of (large) industry within the people (*minkan*), the government should promote the mining industry and bestow machinery upon its people.[11] Neither he nor any other government official suggested, however, that the spirit of mining be awakened within the general population, or that the government should establish model mining and refining operations to serve as a training ground for private mine owners. The basic principle of subsequent mining laws enacted later in the 1870s is indicative of this philosophy: all minerals, with the exception of surface scattered stones used for agriculture, were the property of the government.

Anyone laying claim to, or wanting to open a mine must apply for permission from the Mining Bureau (*Kōzan kyoku*).[12]

The relationship between iron and silk is more complicated than simply a model of heavy versus light industry, or private initiative versus public sponsorship. Economic issues, the need for political validation, issues of 'modernity,' 'civilization,' and ideological materiality all influenced the government's choice of technique and the course of action that would be taken for each industry. Foreign presence in the treaty ports and eventually the interior would also shape the course of action that the government would take in its selection and adoption of foreign technologies. Again, choice of technique was influenced by factors such as ambitious foreign merchants, long-term personal connections with foreign officials, or a belief in a 'civilizational' hierarchy of European countries that are well beyond the scope of simple economic- or politically-based analyses of industrialization.

During the approximately 45-year period covered by this work, Western technological artifacts were embedded and reinvented in terms of their symbolic worth as they became part of Japan's technological landscape. In 1850, the first reverberatory furnace was built in Saga domain as part of *daimyō* Nabeshima Naomasa's attempt at economic and military reform. The new Meiji government would also import foreign technologies with similar intentions, building a *rich nation* and *strong army*, but officials largely relied on an artifact's symbolic value in determining its selection. By 1895, however, the priorities were somehow reversed. *Rich nation, strong army* had become *strong army, rich nation*, and an artifact's ability to symbolically represent 'civilization' was supplanted by perceived demands of national security and military preparedness.

The language and culture of progress

Historians and sociologists of science and technology have long understood the ways that society influences technological development and change. Similarly, they have recognized that technological artifacts have been used to alter socio-cultural patterns. For Meiji Japan, however, I argue that the process was more fluid and interactive. Technological artifacts, or the materials from which the artifacts were constructed, were adopted for their politico-cultural symbolism. The symbolism was then redefined and expressed through the construction of a new set of technological artifacts. In some cases, the new artifacts were the old ones reformulated within a new cultural identity. For this process, I rely on a term of my own making, *cultural materiality*, which is defined as the construction and reconstruction of an identity in terms of technological artifacts. 'An identity' can be individual or group, or even national; and all the members of the group are not necessarily active participants in the process.

In this study, I often follow engineers and examine their proposals while comparing them with contemporary practice. In Meiji Japan, however, bureaucrats and ministers were seemingly at once civil servants and engineers. Despite having limited technical knowledge, or none at all, the perceived urgency of their nation's situation and their collective vision of modernity frequently led them to ignore the technical advice for which they paid so dearly. Their decisions are typically labeled irrational; their ventures as failures. Yet rationality and economic success are not necessarily valid indicators and therefore my analysis of their actions is firmly grounded in an analysis of cultural constructions of material symbolism.

Designing and selecting technological artifacts is not necessarily a formal process based strictly on empiricism or evidentiary reasoning. Technological choice is accompanied by some process of rationalization, that is selection is based on evaluating which characteristics are best suited for attaining one's goals. Moreover, choice can be based on intuition, so-called gut-feeling, social pressure, or physical appearance just as often as it is based on a careful examination of factors such as economic feasibility, an evaluation of skill and labor, or on the availability of natural resources. Historian Robert Friedel once observed that, 'everything is made from something.'[13] Undeniably accurate and simple, this statement goes a long way toward explaining some of the basic premises of this study. I am not especially concerned with the invention of machines per se. I am interested in the materials from which the machines were made and how an artifact's materiality determined its socio-cultural character and selection. While examining the Tomioka mystique and how it was propagated by private reeling firms, for example, I will discuss the simplification of technology and its adaptation to fit the economic realities of the day. This behavior was motivated by utility. Small firms simplified the technologies they observed at Tomioka Filature or elsewhere because they did not have the economic and technical wherewithal to copy the government's factory. The major significance of this behavior, however, is that entrepreneurs used advertising which asserted that their firms were based on Tomioka's technology when nothing could have been further from the truth.[14] The questions then become: Why claim to be something you are not, and what was it about Tomioka that made people choose to be identified with it?

In many ways the machines and the people who designed and used them are read as texts in this work.[15] Technological artifacts are themselves texts, documentary evidence of the beliefs and motives of the people who surround them. As such, they have a degree of independence from their designers and builders.[16] Either implicitly or explicitly, they represent the social and cultural values of the society in which they are located, which may or not be the designer's values. Moreover, these may or may not be the values of the society in which the artifact was originally embedded. A machine's functionality is largely the same regardless of location: whether made from cast iron or wood, reeling frames are used to unwind raw silk filaments from cocoons; a blast furnace with a wooden box bellows smelts iron in much the same way as a blast furnace with a vertical piston blowing engine. And while it is possible to elicit a single meaning from the utilitarian design of the French silk reeling machinery or the massive proportions of the British-built furnaces, my concern also involves the meanings embedded in the basic materials from which the artifacts were made. Why did bureaucrats regularly choose cast iron, steam-powered machinery over the viable wooden, water- or hand-powered alternatives? While at times lacking the word-based character of the historian's traditional documentary evidence, artifacts speak volumes about the societies in which they are deployed.

Material culture, the 'manifestation of culture through material products,' is based on the notion 'that human-made objects reflect, consciously or unconsciously, directly or indirectly, the beliefs of the individuals who commissioned, fabricated, purchased or used them and, by extension, the beliefs of the larger society to which these individuals belonged.'[17] In Meiji Japan both the material representations of 'civilization,' i.e., the artifacts, and the 'beliefs of the larger society,' that is 'Victorian Japan,' were embedded simultaneously.[18] In sum, the assigning of cultural values accompanied the development of the *bunmei kaika* mind set.

This examination of material culture and choice of technique is an agglomeration of ideas related to the ways in which people and cultures assign non-material values to technological artifacts, which in turn reflect these values back onto society. In exploring the invention of the socio-ideological constructions of materials and artifacts, I examine how a society evaluates and adopts a technology for the purpose of re-shaping society. In its most basic sense, this may sound like technological determinism, although it is not. Technological determinism rests on the notion that technologies have an autonomous existence with regard to society. Technological progress is seen as linear; society and social institutions are viewed as having to adapt to technologies. The implication is that technologies are universal.

In the case of Meiji Japan, this was not the case. The mechanization of the iron and silk industries, although infused with the latest Western technologies, did not leap forward as planned. Government-sponsored facilities encountered frequent setbacks despite huge capital expenditures, use of the latest technologies, and well-trained personnel. Mechanization of the silk reeling industry on Western principles took until the turn of the century; and even then it was based on local initiative and the gradual adoption of beneficial technologies judged appropriate to local circumstances.[19] In the iron industry, the most modern technologies could not be universally applied without problems. After nearly two decades, intellectual technology matched experience and the government's 'universal' technology became fully practicable. Although the historical actors within this study tended to believe in the logic of technological determinism and the universal nature of technology,[20] I stress the interplay between society and technology – that neither has an autonomous existence – and the ways in which society affects technology affects society.

In Meiji Japan, a Victorian attitude toward progress shaped technological choice. What Eric Schatzberg calls the 'ideology of technological progress' and by extension the 'progress ideology of metals,' I simply refer to as progress ideology, or progress ideology of materials.[21] Partly because we are concerned with more than metal, and partly because of the frequency with which contemporary Japanese, British, and French writers use the word 'progress,' I have chosen to modify the name, although the theory is largely the same. Progress ideology is the infusion of cultural values into materials deemed 'modern,' accompanied by beliefs in the inevitability of technological progress for the betterment of society. Icons of the first Industrial Revolution, cast iron and steam engines, coal and bricks, were materials that symbolized 'modernity' and 'progress.' Described by words such as 'permanent' and 'sturdy,' a society that had brick buildings and iron bridges, railroad and telegraph networks was 'modern' and 'civilized.' Conversely, wood was considered impermanent and flexible; it was the material of traditional crafts and was presumed to lack the precision of iron and industry. Any society that relied on wood for its buildings was traditional, and that was 'uncivilized.'[22] In selecting the technologies with which to build a nation, the new Meiji government placed almost exclusive reliance on 'modern' materials and the values they embedded.

Meiji officials adopted technologies whose functionality did not match stated expectations. Claiming that the modernization of industry was necessary for economic reasons, government officials rejected out of hand proposals recommending indigenous technologies that were based on evaluations of economic feasibility. Reports clearly stating that it would not be possible to disseminate the chosen technologies were equally ignored. If the government intended to modernize the silk reeling industry in order to disseminate Western reeling technology and to improve the export trade, as was claimed, why disregard recommendations that targeted these goals?

In attempting to mechanize or modernize the silk reeling industry, the Ministries of Finance (*Ōkurashō*) and Civil Affairs (*Minbushō*) first sought technical advice from foreign and Japanese experts. They received proposals for a filature based on the hybridization of Japanese and European silk reeling technologies. The machinery was to be made of wood, steam engines were to be imported from Europe. This proposal was rejected and the foreign adviser was ordered to France where he would purchase cast iron machinery, steam engines, and boilers. The factory buildings were to be constructed from brick. Later revealed in government-sponsored silk reeling manuals was their belief in the superiority of 'modern' materials – their progress ideology. Iron machines and brick buildings were 'modern' and permanent, they were demonstrative of 'civilization' and one's status. Although a reasonable choice of technique if one is trying to import 'civilization,' the government's actions ran counter to their professed, economically motivated goals.

The modernization of Japan's iron industry represents a different type of case study. Again, choice of technique was guided by an ideology of technological progress; the rejection of any artifacts deemed traditional and the unqualified support of the latest European techniques. In this case, the foreign adviser gave the government exactly what it asked for, a large facility based on the most up-to-date methods. Japanese engineers, however, had no experience with operating a large modern ironworks. Moreover, there were a number of technical issues, such as sources of fuel and transportation, that were never fully addressed. While some government advisers urged Meiji officials to be more prudent and to carefully evaluate their choice of technique, the officials proceeded to adopt the latest Western technologies while ignoring the technical and economic ramifications of their decision. And while it will be demonstrated that the Japanese adviser's proposal was rejected for a number of reasons, the most outstanding of which was its reliance on seemingly traditional technologies, it was also inappropriate when evaluated from economic and technical perspectives. In sum, the more reasonable methods were adopted but for the wrong reasons.

As already suggested, one could say that both the materials and ideologies of 'civilization' were adopted simultaneously. If for no other reason this was conditioned by time, or more appropriately, a lack thereof. Driven by the need to maintain national sovereignty, pressured by the foreign powers in the treaty ports, and recognizing the necessity of import substitution, individuals within the Meiji government pushed through their personal agendas of modernization with little thought of the ramifications. Western material culture, first in the form of clothing and accessories, and later in architecture and transportation, was adopted at every level of society. Accompanying the superficial embedding of Western material culture in Japan was the ideology of *bunmei kaika*.

Bunmei kaika was considered the magic bullet that would transform Japan from the backwater of traditional Edo to the cosmopolitan city of modern Tokyo. Accompanying government pronouncements about improving the quality of Japan's raw silk through the adoption of European machines and techniques,[23] or the necessity of building an ironworks to provide for all of Japan's domestic and military iron needs,[24] was the undercurrent of 'civilization' rhetoric. Some believed that anything from the West was 'progress.' One writer in the *Tokyo nichi nichi shinbun* argued that Japan should adopt horse-drawn omnibuses because, although horses tended to scare the Japanese, one finds this mode of transportation in the 'great civilized and enlightened' nations, '*daibunmei daikaika no kuni*' of England and France.[25] Others, like Fukuzawa Yukichi, criticized the superficial adoption of Western artifacts without cultivating the spirit in which they were produced. Fearing that people would try to buy 'civilization,' he cautioned against the propagation of an ideology that would lead the public to believe that Western material culture was 'civilization.'[26] His arguments, however, frequently fell on deaf ears. With time, and perhaps the tacit understanding that Western materiality alone could not 'civilize' Japan, came the gradual rejection of *bunmei kaika* ideology, but not the rejection of the artifacts themselves. Japan still demanded the material trappings of the West, but by the late 1880s these artifacts had become the material trappings of Japan. The superficiality of *bunmei kaika* rhetoric was replaced with more nationalistic sentiments, although the basic terminology used to describe the technologies and the nations upon which to base the comparisons were largely the same.

Following Japan's rapid rise to economic superpower status we tend to think of the Japanese as 'rational people' whose industrial policies are based on long-range planning, purposeful research and development, and strict economic analysis. As a result, studies of Japanese technology rarely look to cultural factors or the perceived value of cultural attributes as influential in designing or adopting technological artifacts.[27] The relationship between technology and culture although implicit, is essential to understanding this study.

The frequency with which words such as 'civilization' and 'enlightenment' or 'progress' and 'modern' appear in this book may pose problems for some readers. Short of inventing new vocabulary to deal with these concepts, I believe it is better to confront them and understand the spirit in which they were and are used. In most instances these words will be used in an historical sense and imply no value

judgement on my part. As historical terms, however, all of the judgmental implications of the historical actors are predicated by their use. In my narrative, when used
in this manner, these words will be in quotes so as to set them off from the present.
To the Victorian world, 'progress' meant the linear advancement of societies and
cultures. It was the process which J. B. Bury so aptly described, and implied a
beneficial outcome.[28] Within the rhetoric of social Darwinism, 'progress' implied
the natural evolution of a society whose evaluation was based on its ranking against
the hierarchy of the self-proclaimed advanced nations.

It was a primary concern of many that Japan 'go forward.'[29] Not everyone
appreciated Japan's willingness to adopt Western materiality. As already mentioned,
Fukuzawa Yukichi and others were critical of Japan's seemingly indiscriminate
acceptance of Western material culture. Foreigners, too, bemoaned Japanese
behavior. It was characterized as 'unsound progress' and many criticized the new
government and its citizens for their willingness to 'throw off in a year the habits,
customs, and institutions which centuries have woven and adapted to the necessities
of the national mind.'[30] Most criticism was couched in the belief that Japanese
'progress' was undirected and undertaken without understanding the true spirit of
'progress' in the West.

'Civilization' in the Victorian vernacular had nothing to do with urbanization or
social stratification. It was what one attained by moving up the ladder of 'progress.'
'Civilization' was measured in a number of ways, most of which were linked to
technical and intellectual achievement and defined according to a European social
model. While becoming 'civilized' countries like Japan at least recognized the
virtues of Christianity, but by and large, science and technology were the basis for
comparing countries.[31] Railroads, bridges, monumental architecture and the like –
again, embodiments of the first Industrial Revolution: steam, iron, and precision –
were hallmarks of 'progress and civilization.'[32] 'Measurements of cranial capacity,
estimates of railway mileage, and the capacity for work, discipline, and marking
time became the decisive criteria by which Europeans judged other cultures and
celebrated the superiority of their own.'[33]

Used in the historical voice, 'modern' is also value-laden. The opposite of
'tradition,' its implications are 'progressive' and 'civilized.' To be 'modern' one
must be scientific and one must recognize the superiority of certain materials – iron,
steel, glass, and brick. 'Modern' meant not only new, but superior. It can be
summarized in the dichotomies of 'modern versus tradition,' 'progress' versus
stagnation, inanimate power versus animate, synthetic versus natural, West versus
East. While quaint, 'tradition' was backward.

Used in the present voice to describe the past, the word 'modern' is used
temporally, only betraying something's relatively recent vintage. When discussing
Japanese technology in terms of temporality, I consider a modern machine one
imported to Japan after Commodore Matthew C. Perry's arrival in 1853. A modern
technique, for example, is one that was recently developed; a traditional practice,
technique, or machine would be one that was familiar and used in Japan prior
to Perry's arrival. It should be noted that traditional techniques in Japan were
not necessarily Japanese. In fact, by the time workers were breaking ground on

Kamaishi ironworks in 1875, a Western-style blast furnace had been embedded in the Japanese industrial landscape for nearly two decades.[34]

Describing the silk industry, I tend to use the word 'mechanization' interchangeably with 'modernization' when discussing recent adaptations of more technically sophisticated mechanisms to traditional Japanese methods. This usage is not without problems because *zaguri*, one of many traditional Japanese reeling machines, is by definition mechanized. My usage, however, is derived from the idea that the imported technologies typically had a greater number of moving parts, e.g., gears and pulleys, and relied on inanimate sources of power. When I use the word 'modernization,' it implies either the recent hybridization of technologies based on the infusion of Western/foreign techniques or mechanisms, or the complete adoption of foreign machinery. That is, modernization or mechanization also speaks of the wholesale adoption of technologies new to Japan. In neither case is there the implication that the more recently acquired Western technologies were superior to the Japanese methods.

My purpose in describing the theories, methodologies, and terms employed throughout this study is to put the reader at ease. Much of the vocabulary of mechanization and industrialization studies is value-laden and highly subjective at best. This is, in fact, reflected in the literature of Japanese industrialization. Authors regularly divide into two camps: indigenous technology as the primary contributor in reshaping the Japanese industrial landscape; or proponents of government-led initiatives to import Western technology. Stemming from this bifurcation is a defensive response by many scholars in Japan to theories which seem to 'challenge' the worth of traditional technologies.'[35]

My intention is not to evaluate the functional worth of Japanese technology. Rather, I examine the historical actors' arguments that surrounded the adoption of a technology, and situate them within contemporary practice in an effort to test the validity of their claims. By doing so, I am able to establish that their actions were guided by an ideology of technological progress. Recognizing that the historical actors who formed the Meiji technological decision-making community had individual agendas, as do the authors who write about them, is crucial to understanding how people choose technologies and the nexus between ideology and artifact.

Meiji modernization revisited

Iron and silk share a long proto-industrial history in Japan. Given that many of the nineteenth-century changes to the iron and silk industries were profound, it is necessary to understand where these industries began and to examine some of the modifications that these traditional technologies underwent prior to the introduction of Western methods. Because many of the traditional methods continued to prosper until the end of the nineteenth century, it is also important to know what they were. This is the subject of Chapter 2. Changes to the iron and silk industries took place within the larger scheme of Meiji industrialization and were subject to the constraints – real, perceived, and artificial – imposed on the government's overall plan, *shokusan kōgyō*, the so-called program of industrialization. It is, therefore, necessary to examine briefly the general history of Meiji industrial modernization. This too, is the subject of the second chapter.

Silk was the first trade commodity to be exported from Japan on a large-scale basis. By the mid-1850s, much of Europe's sericulture and many of its silk reeling industries had been devastated by the pébrine virus, which prevents silkworms from spinning their cocoons or kills them outright. Without a supply of cocoons, trading houses and manufacturers alike were forced to look elsewhere for the basic raw materials of the silk trade: raw silk, cocoons, and silkworm egg cards. China rapidly became a source of raw silk for the European market. Political turmoil in the form of the Taiping Rebellion (1850–64), however, led many European trading houses to search elsewhere. What spelled possible economic catastrophe for China was opportunity for Japan as French, British, and Italian trading firms quickly established themselves in the treaty port of Yokohama. Maintaining a steady flow of high-quality raw silk became a task of primary importance to the Yokohama merchant community.

Seemingly as quickly as the demand for, and exports of, Japanese raw silk increased, the quality, conversely, declined. In an unprecedented action aimed at remedying the situation, British merchants represented by the Yokohama General Chamber of Commerce successfully petitioned Sir Harry Parkes to present public works minister Date Munenari with a letter outlining both British frustrations and details for possible solutions to the quality problem. Sir Harry considered the matter to be quite serious and sought to impress this fact on the Japanese bureaucracy.[36] Minister Date too, took the matter seriously, having the petition and outline of suggested improvements translated into Japanese for dissemination to the silk producing community. He also thanked the British for their concerns and explained that the government was already undertaking efforts to remedy the situation.[37]

This was not simply a matter of British and Japanese officials placating angry merchants. London was the world's hub for the transshipment of raw silk. Nearly all the silk destined for Lyons, Marseilles, and the ports of Italy passed through London. Japanese merchants had recently taken to selling their best silkworm egg cards to French and Italian merchants who hoped to bypass the London market in an attempt to ensure the recovery of Europe's sericulture industry. Myopically, it made economic sense for Japanese merchants to sell off their best quality silkworm eggs and unreeled cocoons. They would reap high profits without any of the expenses of rearing the silkworms or reeling the silk. That this practice would endanger Japan's raw silk export market did not seem to matter at the time.[38] The British, however, refused to allow 'merchant greed' to interfere with their plans and acted accordingly by pressuring the Meiji government to stop the outflow of silkworm egg cards and urge the adoption of Western reeling techniques.

French merchants took a more discreet approach. Rather than trying to bully the Meiji government into action, they offered to help (the government) build a model reeling facility to train Japanese reelers in Western methods.[39] Although initially rejected by officials within the Meiji bureaucracy like Itō Hirobumi, this was the impetus behind Tomioka Silk Filature.[40] But Tomioka would not be built on the basis of economic or technical assessments that focused solely on improving the quality of Japan's raw silk. At the time Tomioka was conceived and constructed, some domain officials had been building their own filatures based on imported Western technology. If the new Meiji government was going to build a model filature, it would be multi-functional. First and foremost, Tomioka would be built to situate the Imperial government above the domains which were still vying for political influence. Second, it would serve as a vehicle for the importation of Western 'civilization' to Japan. Third, but not of least importance, Tomioka would aid in improving the quality of Japanese raw silk through the dissemination of Western methods.

Chapter 3 shows that Tomioka's physical presence corresponded more to perceived ideals of 'enlightenment' and 'civilization,' as evidenced in the choice of materials for the machinery, buildings, and fuel, than it did to the technical and economic advice the government received from its own advisers – foreign or Japanese. I examine the government's plans for Tomioka and contrast these with their adviser's proposals, demonstrating that Meiji officials abandoned viable alternative technologies because they were too 'traditional' and would have done nothing to elevate the government's prestige vis-à-vis the domains or in the eyes of the Western powers. Their selection of building materials is revealing. Brick was used for buildings – when brick was practically non-existent in Japan – because stone and brick were considered indicators of a country's level of civilization according to Victorian standards. Cast iron was deemed essential for the machinery because iron, not wood, was the material of 'modern' nations. Similarly, coal had to be used for fuel because likewise, 'modern' nations used coal, not kindling fires. These choices reveal the ulterior, and guiding motivations behind the construction of this 'model' facility.

Through an examination of publicly sponsored and privately written silk reeling manuals, I further demonstrate how viable alternative technologies were disregarded because of their inability to be identified with 'progress' and 'civilization.' In addition to betraying a sense of *cultural materiality*, these manuals, which typically recommend the adoption of French reeling methods over Italian, illustrate a belief in a particular European civilizational hierarchy. Internationally, Italian methods were simpler and considered superior. They were also economically more viable and easier to master. For many Meiji bureaucrats, however, Italy lacked France's prestige and as such, the latter's technology was imported by the government and widely recommended for private use in government sponsored silk reeling manuals.

In the end, Tomioka would bring prestige to the new Meiji government and serve as an exemplar of Western 'civilization' in Japan far longer and better than it ever functioned as a model filature or as a business venture. As the government's yardstick of 'civilization and enlightenment,' Tomioka and its silk would serve as the basis of comparison by which Meiji bureaucrats elevated Japan above China and even some European countries. Although Tomioka Silk Filature was dedicated as a model factory whose purpose was to disseminate modern mechanized reeling techniques and train workers in these methods, its true function was political legitimization through the ideological construction of cultural values based on its technological artifacts.

Developing a modern iron industry also relied on the importation of foreign technology and its accompanying ideological baggage. This is the subject of Chapter 4. Concomitant to the public works ministry's policy of railroad development, Meiji officials built Kamaishi Ironworks with the unrealistic goal of satisfying all of the new nation's demands for iron. Envisioned as the Krupp Ironworks of Japan, Kamaishi would be unequaled in magnitude.[41] Compared to the former shogunate's ironworks at the Nagasaki Shipyard, Kamaishi was larger and more modern. Compared to anything the domains had constructed in the years prior to the Meiji Restoration, the new government's facility was a techno-behemoth. In this sense, Kamaishi had the potential for similarity to Tomioka. As a modern industrial icon of the West, Kamaishi could also help situate the new government above the former domains and shogunate. Kamaishi, however, was not destined to symbolize Japanese 'progress.'

From the start, the government's new ironworks was controversial. Foreign governments, the British in particular, had little or no interest in having Japan develop its own iron industry. Having been legislatively excluded from extracting Japan's mineral wealth, British officials made known their displeasure and turned their backs on the Meiji government's efforts at building a modern iron industry. In fact, the British looked to Japan as an export iron market in the face of declining iron prices worldwide. The Japanese government's foreign advisers also recommended against the project. At least two advisers considered the economics of the issue: importing British iron was less expensive than erecting an ironworks for domestic production. Others based their views on more technical perspectives: there were no Japanese familiar with the operation of large modern blast furnaces. Arguing that Japan must not be reliant on foreign iron for 'civilization' building, however, Meiji officials mostly turned a deaf ear toward their advisers' economic and technical warnings.

Ignoring advice does not mean that it went unheard, however. Knowing that foreign governments and advisers alike doubted the feasibility and future success of Kamaishi Ironworks, Meiji leaders were cautious in their public appraisal of the facility's ideological worth. Although embedded within the discourse of 'civilization and enlightenment' as a 'modern' material, iron was only one component in the process. It was not just the iron of an iron bridge or iron machinery, for example, that was symbolic of 'progress'; it was the strength, permanence, and precision embodied within these structures that made them icons of Western 'civilization.' As such, Kamaishi was a component of 'civilization' building, but it was one that could be considered unnecessary given Japan's ability to import iron from abroad. This was not lost on Meiji officials who wanted to break free of their dependency on British iron while simultaneously making progress toward achieving 'civilization.' Unlike the government's filature, there was little fanfare or public recognition of Kamaishi Ironworks. No one ever proclaimed the ways in which Kamaishi Ironworks represented 'progress,' as had been the case with Tomioka's silk. If the Kamaishi project was a failure, the setback would be largely economic. Japan would not suffer a blow to its national prestige.

Throughout the period of Kamaishi's construction, roughly 1875 through 1880, foreign observers in the treaty ports alternately lavished praise and criticism on

Japan and the Meiji government for the country's 'progress' and willingness to embrace Western material culture. As time passed, the voices of criticism became more frequent and the tenor of attacks more pitched. Japan was accused of mimicry and not understanding the intellectual basis of Western 'civilization.' As the target of much of this criticism, Meiji officials maintained public silence when it came to Kamaishi. Unlike other projects aimed at bringing Western 'civilization' to Japan, such as Tomioka Silk Filature, building a railroad system, or a telegraph network, the newspapers did not report about the events in Iwate prefecture. If Kamaishi was a success, it would challenge foreign allegations of Japanese apery; if it failed, Kamaishi would provide support for the foreigner's increasingly hostile claims.

Kamaishi's designers and engineers were successful in transferring the latest British iron smelting technologies to Japan. The ironworks sported the latest advances in Western smelting technology: large efficient blast furnaces, stoves, boilers, and engines, plus all the appurtenances of a full-blown refinery. Steam locomotives provided the power for Kamaishi's internal railway, and a telegraph line provided for rapid communications. For all their efforts, however, Kamaishi's engineers were unable to keep the facility in-blast.[42] An accidental fire one cold December morning claimed nearly all of Kamaishi's fuel reserves, and the facility was shut down. After a mandatory government inspection and a decision to secure more fuel, the facility was put back into blast, and once again Kamaishi was producing high-quality pig iron. But Kamaishi's new fuel was of poor quality and the furnaces choked and chilled within six months. This time, however, there would be no debate; the Meiji government abandoned the facility. Kamaishi's blast furnaces would remain cold for nearly a decade.

Throughout Kamaishi's trials, the government maintained its silent vigil. There was no extended debate, no minister or ministry spoke on behalf of the damaged ironworks. It was deemed best not to draw attention to the government's foibles and lend credence to foreign criticisms of the superficial nature of Japanese 'modernization.' Sold quickly as part of finance minister Matsukata Masayoshi's deflationary program, the government's ironworks fell victim to competing ideologies. A self-proclaimed proponent of laissez-faire policies, Matsukata sold Kamaishi for less than half of 1 percent of what it cost to build.[43]

The government's ironworks also lacked the political value of its filature. Internationally, no foreign power was interested in seeing Japan become a self-sufficient iron producer. By 1883, the foreign powers had long since recognized the Meiji government as the legitimate heir to Tokugawa political authority. Political legitimization through the maintenance of expensive icons of Victorian material culture was no longer necessary. This understanding was also reflected through the domestic political scene. The former domains had been 'returned' to the emperor in late 1871, and the last major civil uprising was put down in 1877. Some Meiji officials had even begun to entertain ideas of popular participation in government. In short, the Meiji government was the recognized and established political authority whose physical manifestations of authority were rapidly being replaced by its imperialistic adventures on the Korean peninsula.

Japanese efforts to modernize industry were part of the so-called movement to attain 'civilization and enlightenment.' Often considered an outgrowth of the 1868 Imperial Charter Oath which stated that 'knowledge shall be sought throughout the world,' the slogan *bunmei kaika* provided justification for nearly every social, cultural, political, and economic change enacted during the first two decades of Meiji. With the return of the Iwakura Mission in 1873, official efforts and popular acceptance of 'modernization' only increased. By 1875, it seemed that nearly every corner of Japanese civilization had been touched by the 'Westernizers.'[44] *Bunmei kaika* had two general components: ideological and physical. The former aspect dealt primarily with the promotion of Western enlightenment thought: liberalism, positivism, materialism, and utilitarianism, which were randomly absorbed in Japanese intellectual circles.[45] While these two elements did not necessarily exist in mutual isolation, it is the latter component of *bunmei kaika*, the physical manifestations of 'civilization,' with which we are more concerned; and this is the subject of Chapter 5.

In the simplest sense, physical manifestations of Western 'civilization' ran the gamut from some of the Meiji government's most arduous efforts, such as the promotion and building of a railroad network, or rebuilding the Ginza in brick, to the most superficial efforts by both the government and general population, such as the adoption of Western-style dress, haircuts, and food, or regulations which stipulated that government officials must wear Western formal attire when conducting the business of state.[46] Critics of civilizational superficiality, whether foreign or Japanese, argued that the government's ever-changing policies made a mockery of 'civilization' building. Japanese critics like Nishimura Shigeki considered a lack of moral guidance as the cause of the problem.[47] Foreigners did not necessarily consider this a moral issue. For them it was a problem with the Japanese 'character.' Many resented the Meiji government for trying to 'modernize' the 'new nation' too quickly, while not absorbing the 'spirit' of Western 'progress.'[48] Whether Japanese or foreign, however, the critics overlooked the ideological basis of many of the government's programs. In utilizing the material icons of Western culture, the government was neither attempting to import liberal or utilitarian values to Japan, nor foster Confucian moralism. Meiji leaders thought that they had identified the elements of European and American society that set them apart from the rest of the world. Moreover, they had 'discovered' the yardstick by which to measure the hierarchy of Western 'civilizations.'

Whether at international exhibitions or on study tours, men like Sugiura Yuzuru and Shibusawa Eiichi, who had traveled to Europe as part of Tokugawa missions, or Itō Hirobumi and Ōkubo Toshimichi, who toured the West before and after the Meiji Restoration, recognized that European countries were evaluated quantitatively by their technological achievements. At the 1867 Paris Exhibition, for example, Shibusawa observed that 'by looking at the differences between the countries' exhibits, one can gain a general idea of each countries' customs and level of intelligence.'[49] Shibusawa continued by noting that even at the most basic level, through an examination of clothing and simple utensils, one could see Western superiority.[50] Ōkubo similarly observed on his tour that steam-powered industry was the basis of British might (and that had only come about in the last 50 years).[51] Upon his return to Japan, Ōkubo lost no time in promoting Japanese industrialization as a means through which to gain political parity with the Western powers.[52]

To men like Shibusawa or Ōkubo, these icons of the West were 'civilization.' Following that assumption, importing Western material culture to Japan meant embedding the values of Western 'civilization' in Japan. Through this process, Japan would in essence become a 'Western power.' Whereas modernization did not necessarily mean Westernization, in the case of early Meiji Japan this was at least a possibility.[53] Some extreme proponents of *bunmei kaika* even advocated 'remak[ing] the Japanese in the image of the European, inside and out, even though it would take conversion to Christianity or miscegenation or both to do it.'[54]

More level-headed proponents of *bunmei kaika*, like Sano Tsunetami, took a different approach to building 'civilization' in Japan. Recognizing that international and national exhibitions were catalogues of 'enlightenment,' places where countries competed to display national 'progress,' the Meiji government appropriated the tools that Europeans and Americans used to differentiate their 'civilizational hierarchy' by supporting Japan's participation in, and developing its own, exhibitions.[55] Based on the belief that everything related to these displays represented 'modernity,' the government used the exhibitions as opportunities to promote models of industry.[56] Even before the government officially declared its position in favor of the national and international exhibitions, foreign observers were confirming what Meiji leaders knew all along. Commenting on a newly opened exhibit at a former Tokyo Confucian academy, British observers stated that:

> the formation of a collection of this kind [curiosities of nature and art] is usually characteristic of an advanced stage of culture; and in imitating the European and American example in this respect the Japanese show their great superiority to the Chinese and other Oriental nations.[57]

Tomioka Silk Filature's construction coincided with the Meiji government's official entrance into the world of international exhibitions. In 1873, its silk won medals of honor in Vienna, and international recognition of the nation's 'progress.' Flushed with pride, Tomioka became a symbol of Japanese 'progress' and proof of 'civilization and enlightenment.' Shortly thereafter, various ministries within the Meiji government began promoting local, regional, and national exhibitions – 'mutual-progress exhibitions' – or *kyōshinkai*. More than promoting the dissemination of Tomioka's reeling methods, however, these exhibitions served to promote the Tomioka mystique.[58] Regardless of technique, many private producers claimed at least some association with Tomioka's methods. Participation in these exhibitions also had more objective rewards. Having one's product recognized brought prestige to the manufacturer and a high price for its goods, again regardless of technique. The mutual-progress exhibitions helped commodify *bunmei kaika* for tangible economic reward.

Although briefly mentioned in the catalogues of the 1876 and 1878 International Exhibitions in Philadelphia and Paris, respectively, Kamaishi remained outside the venue of these displays of 'civilization.' This lack of attention was, in part, because the government tended only to publically promote the production of copper, silver, and gold. But it was also a function of changing Japanese attitudes toward the rampant adoption of Western material culture and a heightened rejection of *bunmei kaika*. By the time Kamaishi came on-line in 1880, many in Japan had already begun to question the efficacy of indiscriminately adopting the icons of Western culture. Constant critics, such as Motoda Eifu, who served as Confucian lecturer to the Meiji emperor, became the vanguard of a conservative backlash.[59] Former 'enlightenment' supporters like Itō Hirobumi also withdrew their support and similarly began to endorse the conservative cause.[60] Rejecting the universality of *bunmei kaika* ideology, the belief that 'progress' was part of some historical developmental pattern, other conservative ideologues promoted a Japanese nationalist course of action. The demise of Kamaishi as the national ironworks corresponded with the demise of *bunmei kaika*. And although contradictory policies were not unusual for the Meiji government, it was difficult to promote 'Japan's distinctive national ethics' while simultaneously embracing and embedding Western cultural icons into Japanese society.[61] As Japan turned from cultural assimilation toward self-assertion, even Tomioka eventually succumbed to the new ideology.[62]

Rejecting cultural materiality did not mean abandoning the ideological basis of technological determinism in a world of 'modernizing' nations. The 1880s and 1890s were times of growing European nationalist sentiment and surging imperialism. It is not that imperialism and colonialism had not existed before, after all, these 'isms' were part of Japan's impetus to 'modernize' in the first place. But the tenor of European and American empire building had risen sharply.[63] Prior to the 1880s, Japanese foreign activity can be read as an extension of domestic policy where Meiji leaders sought to consolidate political power and better define national boundaries.[64] Entering the second decade of the Meiji era with greater assurance of its domestic situation and political abilities, more conservative Meiji leaders embraced the basic ideals of 'civilizational' superiority as they increasingly tested Japan's mettle in the international arena.

Accompanying the rise of Japanese nationalism, in part promoted through efforts to topple superficial *bunmei kaika* materiality, Meiji leaders began to formulate their own plans for Japanese imperialist ventures in Korea. Spurred by popular sentiment and a fear of appearing weak after a number of Japanese military advisers to the Korean government were killed in an 1882 coup, Japan succeeded in extracting apologies from the Korean government. The incident, however, solidified the realization that Japan would need to build a larger, more modern navy, especially because it feared that China had pulled ahead in this regard. At the time, however, Japan's financial basis was anything but assured as the country was in the throes of runaway economic inflation. Its efforts to develop a 'modern' iron industry for just this purpose had seemingly failed.

Rather than continue to support the Kamaishi Ironworks, the Meiji government opted to sell it and import the iron it needed for building its navy. A move Itō Hirobumi later recognized as short-sighted, selling Kamaishi was economically and ideologically acceptable at the time.[65] Financially, Kamaishi was a large strain on the government's budget. On one ideological level, heavy industry, or at least non-agricultural industries, were not considered worthy of the state's efforts and should be left to the private sector according to finance minister Matsukata's so-called laissez-faire policies. On a more basic level, however, Kamaishi had become unnecessary. It never had Tomioka's symbolic value, though it might have acquired this in the event of greater success. More importantly, the government had already achieved many of the goals for which Kamaishi was constructed: the creation of a 'modern' nation as defined by Victorian values.

In the end, *bunmei kaika*, Japan's embracing of the ideological and cultural artifacts of the West, would be supplanted by conscious efforts to adopt a policy more closely fitting Maquet's definition of 'modernization,' that is not Westernization.[66] Late Meiji ideologues, while borrowing the logic and actions of Western imperialists, would increasingly couch their arguments in terms of cultural particularity, that is, being 'Japanese.' With Japan having already established its technological ranking among the 'modern' nations of the world, the government was ready to move on to the next historical stage of its development: imperialism and colonization.

2 Tradition and modernization

Before examining the changes that entrepreneurs and Meiji government officials sought to impose on the iron and silk industries, it is perhaps best to look at the traditional technologies, or at least the state of silk reeling and iron smelting technologies on the eve of transformation. After, we will turn to a brief examination of state and private efforts to modernize Japan's industries.

Silk reeling and re-reeling

Sericulture and its associated technologies are said to have arrived in Japan by the third century from China. Silk quickly found favor in the imperial household and stories abound as to which emperors and empresses aided in the spread of sericulture and silk throughout Japan. Silk reeling, the process by which silkworm cocoons are turned into thread, typically involved three distinct stages: unwinding the outer covering of the cocoon, known as *noshi*, to produce a low-grade, impure silk thread; reeling the pure silk; and re-reeling, a process that further refines and strengthens the skeins of silk thread.[1]

Typical reeling apparatus consisted of a brazier and a water-filled vat that were set near some type of device used to unwind the silk. The earliest reeling devices, which date from at least the fifteenth century, *doguri*, were cylindrical pieces of wood suspended horizontally on two wooden legs. The reeler gathered numerous silk threads, twisted them together with her fingers, and either passed the threads through a loop of hair or let them ride over her fingers while winding them onto the cylinder. To turn the cylinder, the reeler struck/pushed the cylinder with the palm of her hand. An improvement to *doguri* was *tebiki*. Here, the wooden cylinder was replaced by a four-spoked wooden frame that was supported on vertical uprights by an axle. The reeling process was basically the same as with *doguri*, except that the reeler turned the *tebiki* by means of a handle at the end of the axle. As with *doguri*, thread tension was maintained by passing it over the reeler's fingers or through a small loop of hair located in front of the take-up wheel.

Beginning in the 1750s, several changes occurred in reeling technology that would increase productivity and improve the uniformity of the raw silk thread. One was the introduction of the *zaguri*, a new machine which was basically one or more four-spoked, horizontal take-up reels, approximately eight inches in diameter, driven

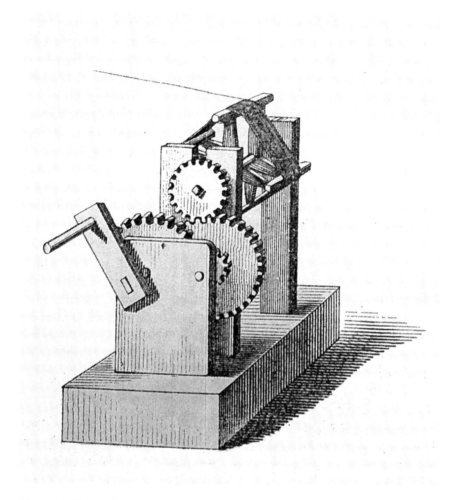

Figure 2.1   Zaguri (*Source*: de Bavier, 1874: n.p.)

by a series of cogs connected to the hand crank. There was a fairlead-like device in front of the take-up reel, the function of which was to feed the thread onto the reel in a criss-crossing pattern. Below the fairlead was a small arm to which was tied a loop of the reeler's hair. *Zaguri* spread throughout Japan's silk reeling districts with surprising speed; there were a number of variations such as ones where the cog wheels were replaced by drive belts. The *zaguri* and its precursors were made completely of wood, bamboo, and fiber-based cord.

To begin reeling, a reeler, most often a young woman, would place several cocoons into the iron vat of hot water. As the hot water softened the natural adhesive that bound the cocoon, the reeler would gently stir the cocoons with a pair of sticks,

Figure 2.2 Woman reeling silk with belt-driven *zaguri* (*Source*: de Bavier, 1874: n.p.)

often made of willow, to help free up the ends of the silk filaments. The reeler would next gather the free filaments with the sticks and then roll them between her fingers forming a thread. She would place the thread onto the take-up reel and begin turning the machine's handle, drawing the silk from the hot water bath onto the reel.

After several turns of the handle, the more coarse *noshi* would be reeled off the cocoons exposing the more valuable pure silk. At this point, the thread was cut, the ends of the pure silk set aside, and the process repeated with new cocoons until the take-up reel was full of *noshi* and there were enough cocoons in the bowl to reel a large quantity of pure silk.

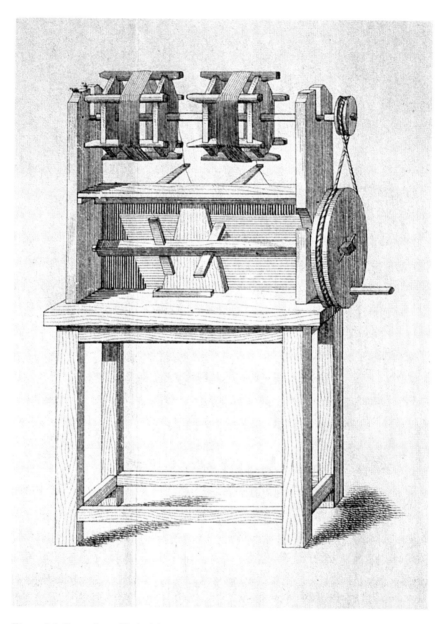

Figure 2.3 Rear view of belt-driven *zaguri* (*Source*: de Bavier, 1874: n.p.)

Depending on the denier of the silk thread being reeled, the woman would take between 8 and 13 silk filaments and twist them into a thread with her fingers. She would lay the thread over a smooth stick that was placed on the edge of the bowl, parallel to the *zaguri*, pass the thread through the loop of hair, feed it over the fairlead, and tie it onto the reel. She would then begin turning the *zaguri's* handle, drawing pure silk thread onto the reel. If a filament would break, the reeler would find a free filament and toss it onto the moving thread, restoring it to the desired denier. By running the thread over the stick and through the loop of hair, excess water was stripped from the thread while the friction helped to bind the individual filaments more firmly.

Before the silk could be sold it needed to be re-reeled into a larger skein and dried. This process was a necessity with Japan's humid climate. Damp silk left on small reels in humid areas tended to coagulate into a sticky mass. Re-reeling allowed the silk to dry more efficiently as well as attain a more uniform texture. The device used to re-reel the silk was much like the *zaguri* except that it was much larger in size: its reels being approximately 16 inches (41 cm) in diameter. There was no fairlead, but instead a series of semicircular wire hoops, typically 10 or 12, attached below the reel. A *zaguri* was placed below the re-reeling machine's take-up reel, directly under a wire hoop. The silk thread was run over the wire and tied to the reel. Once the operator began turning the re-reeling machine's handle, silk was transferred from the *zaguri* to the larger reel. By passing through the semicircular wire hoop, the thread was not spread out over the reel, as was the case with the *zaguri*, and the skein was kept to a relatively compact size. Once the re-reeling operation was completed, the skeins of raw silk were removed from the reels and allowed to dry. Once dry, they were ready to be graded and packaged for sale.

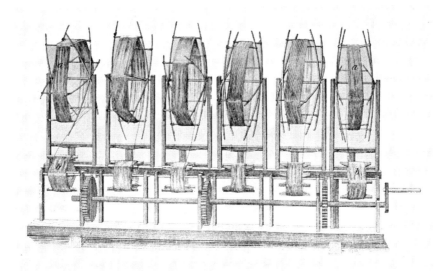

Figure 2.4 Machine used for re-reeling silk (*Source*: de Bavier, 1874: n.p.)

Early iron smelting and mining techniques

The technologies associated with iron smelting first appeared in Japan in the northern part of Kyushu and developed from the third through first centuries BCE. By the Yayoi Period (50–250 CE) iron tools became more plentiful, as is evidenced by advances in woodworking technologies. By the last century of the Yayoi, iron-working technologies spread quickly across the central region of Japan from west to east. Over the course of the next several hundred years, iron completely replaced stone as the mineral of choice.[2] Iron swords, armor, and arrowheads along with axes, adze heads, chisels, woodworking knives, and sickles came to occupy prominent places in the tombs of the Kofun period.[3] From that time onward, iron and its alloy with carbon, steel, were Japan's pre-eminent proto–industrial metals.[4]

The early Japanese iron industry typically relied on the use of iron sand which was extracted since medieval times from surface-dug conical pits.[5] Several volumes of historical literature, such as the *Nihon Ryoiki* and the *Konjaku Monogatari*, mention this type of mining, as well as the labor involved in sorting and classifying iron.[6] In 1610, open pit mining was abandoned for a more economical alternative, the water method, known as *kanna nagashi*. This method appears to have been first adopted in the Chūgoku mountain range of western Japan.[7] The method involved choosing a site with a steep rock face and directing a stream of water to flow down the face through a specially prepared conduit laid in a trench. A typical conduit measured approximately 14 inches (36 cm) deep, 29 inches (75 cm) wide at the top, and 19 inches (48 cm) wide on the bottom.[8] Stones, rocks, and earth were raked into the rushing waters and carried downhill. The action of the water would separate the earth and stone from the iron sand, which would settle in a pond prepared at the bottom of the conduit. Because different grades of iron sand have different specific gravities, the water also took care of separation. This was truly a labor-saving device.

Whereas water was a blessing for these early miners, it was a curse for later miners who dug for iron ore. The iron mines of the late Tokugawa era have been compared to Roman and Carthaginian mines because of their haphazard layout. A typical mine was a series of long twisted galleries that seemed to go wherever the digging was easiest or most profitable. The biggest problem, however, was water. The miners, having only hand pumps, were often forced to abandon the mines when seepage became excessive. The amount of water must have been serious; the name given to the lower levels of the mine was *mizunuki* or drain.[9] Other factors that increased the difficulty of working these mines were the lack of mechanical devices to help bring the ore to the surface and the absence of any ventilation except for the entrance.

Although there were alternatives to surface mining in pre-modern Japan, iron sand was the preferred raw material of the iron industry. There are two varieties of iron sand in Japan – black and red. Black iron sand, called *masa kogane*, is pure magnetite and has a low phosphorus content. Red iron sand, called *akome*, contains ferric oxide and other impurities such as silica and titanium oxide.[10] Both were the end products of the natural erosion of granite and andesite found in the mountainous parts of the country.[11] It appears to have been the preferred raw material because it was simple to work and required less labor than did separating iron ore from rock.[12]

Smelting iron sand took place in *tatara* furnaces which existed in Japan in various forms since approximately 300 BCE.[13] The earliest type of Japanese furnace used for iron smelting is called *nodatara*, literally, 'furnace in a field.' It took eight to ten men to operate the bellows that were on either side of the furnace. The *nodatara* was only used for approximately 100 days a year; from the vernal equinox to the autumnal equinox.[14] The next noticeable form was called *tatara*. It differed in that it was put in a building called a *takadano*.[15] This allowed the smelting process to go on year-round. The furnace itself is a V-shaped trough made of clay. *Tatara* ranged in size from approximately 5.5 to 15 feet (1.7 to 4.8 metres) in length and 2.25 to 6 feet (0.7 to 1.8 metres) in width. Most were approximately 3 feet (91 cm) deep and had side walls that were 8 or 12 inches (20 or 30 cm) thick at the upper edge.[16] The inside of the furnace was lined with a refractory clay that acted as a flux during the smelting operation. As a result, the furnace was destroyed by the smelting process. *Tatara* also had a series of holes for tuyères along the bottom edges, a single tap hole at one end, and an inspection hole or *hodoana* through which the ironmaster or *murage* could keep watch on conditions inside the furnace.

To smelt iron, iron sand and charcoal were alternately layered in the *tatara* and it was lit. Once the furnace grew sufficiently hot, the red sand, *akome*, was quickly reduced while absorbing the carbon from the charcoal. This was tapped from the hole at the bottom of the furnace producing pig iron or *sentetsu*. Once the furnace temperature stabilized, black sand, *masa kogane*, was added and slowly reduced over a 2 to 3-day period.[17] Throughout the smelting operation, charcoal was added to the furnace and laborers continuously worked the bellows at the direction of the ironmaster. Once the ironmaster decided that the process was complete, the ends of the *tatara* were broken off and the fire raked out. Furnace yield varied by size and quality of materials. A *tatara* found in the village of Hinosako, for example, used approximately 7.5 US tons (6.8 tonnes) of charcoal and 9.4 tons (8.53 tonnes) of iron sand to produce approximately 5 tons (4.54 tonnes) of iron and steel.[18]

At the end of the smelting process, a lump of steel and iron called *kera* remained. If black sand was used it would contain hyper-eutectoid steel (carbon content 1.2–1.7 percent) called *tama hagane* and pieces of iron with a lower carbon content (less than 0.8 percent).[19] The *tama hagane* was the first quality steel used in swords. The pig iron drawn off earlier in the process was used for casting and the iron that formed around the *kera* or a *kera* that was made from red sand, called 'knife iron,' was worked by blacksmiths into wrought iron for farm tools and other daily use items.

A significant innovation in early Japanese iron smelting technology was the evolution of the bellows. Original Japanese bellows were similar in design to Chinese double-piston bellows that allowed a constant flow of air to be directed into the furnace. There were usually eight or ten of these devices, four or five to a side, around the *tatara* requiring just as many laborers. Once the *tatara* was moved indoors, the *fumifuigo* was developed and the *tatara* was officially known as *fumidatara*. A *fumifuigo* is a foot-operated bellows that only reduced the number of workers slightly, if at all, but increased the air flow to the furnace. The greatest innovation came in 1691 with the invention of the balance bellows, *tenbinfuigo* or *tenbindatara*, as the entire furnace–bellows assembly came to be called, that required only two men for operation. Greater activity in the iron industry is associated with its introduction.[20] In addition to decreased labor requirements, the *tenbindatara* permitted the attainment of higher furnace temperatures necessary for smelting cast iron, and gave the ironmaster the ability to decarburize iron making it into steel.

The most obvious problem with the *tatara*, or any of its variations, was that the furnace was destroyed with each use. And although *tatara* were in use well into the late nineteenth century, they were only suitable for small-scale iron production. In addition, the quality of the iron produced in *tatara* was unequal. This problem would come to the fore in 1855, when Nambu *han* technologists would try to convert iron smelted from iron sand in a *tatara* in a recently constructed Western-style reverberatory furnace. The inferiority of the product forced the development of alternative means and raw materials, i.e., iron ore, for smelting pig iron.

As will be revealed in the following chapters, much of the impetus for Bakumatsu and early Meiji era technological change in the silk and iron industries came from the realization that some traditional Japanese technologies were ill-suited for large-scale industrialization. Part of the problem faced by these traditional industries was the irregular quality of reeled silk and pig iron after entrepreneurs attempted to increase production to meet the demands of the export market for silk, and the Meiji government's demands for iron. Although Meiji officials rapidly embraced Western methods as the basis for Japanese industrialization, choice of technique and subsequent transfer of technologies relied more on perceived cultural values than utility.

Industrialization modernization: an overview

Meiji industrial modernization is most often viewed from one of three perspectives: emphasis is placed either on state-sponsored industrialization prior to 1885; or on the 'take-off' years of private enterprise and Japan's industrial revolution following 1887; or on the contributions of traditional industry and human capital. While it is without doubt that the Meiji government was pivotal in its efforts to foster the growth of strategic industries and build national transportation and communications infrastructures, one cannot ignore local efforts at industrial modernization. It is with these perspectives in mind, state-centered, private initiative, and the role of tradition, that we will survey the history of early Meiji era industrialization efforts by examining several representative industries. Because Chapters 3 and 4 are case studies of the silk reeling and iron industries, we will not be discussing them here. It goes without saying, however, that as Japan's most valuable export commodity throughout the nineteenth century, and as the material from which to build a nation, silk and iron respectively, are significant to the overall story of industrial modernization.

Prior to the Meiji Restoration, Western technical knowledge was imported by the Tokugawa *bakufu* and by some *daimyō*, many of whom were largely autonomous of *bakufu* control. After the collapse of the Tokugawa *bakufu*, the Meiji government became the primary importer of foreign industrial knowledge. As will be discussed in the next chapter, some of the *daimyō* domains and merchant houses also actively sought foreign knowledge, if only to a lesser extent.[21] The new government's first move was to take over arsenals and factories established by the Tokugawa regime and then to continue its focus on strategic and heavy industries. Its efforts to build a modern, industrial nation came under the guise of a policy known as *shokusan kōgyō*, literally 'increase production and promote industry.'[22]

Like other early Meiji policies and activities, *shokusan kōgyō* did not exist in a vacuum; it was a program established alongside *fukoku kyōhei* (Rich Nation, Strong Army). In short, Meiji industrial policy was designed to help the country achieve its goals of building an economically and militarily strong nation. As a result, two ministries were created whose charge was to facilitate the absorption of modern Western industrial technologies. The Ministry of Public Works was established in 1870, three years later the Ministry of Home Affairs (*Naimushō*) was created with several overlapping jurisdictions and the common goal of helping to coordinate Japan's industrialization.

From its inception the Ministry of Public Works was charged with directly importing or facilitating the importation of Western technologies, building strategic and heavy industry, and with providing an educational basis from which Japan could continue its industrial development. A look at the ministry's charter is instructive. The Ministry of Public Works was established to make Western technology more accessible and understandable to the people. It was to manage the nation's natural resources, which included opening and overseeing mines; establishing metallurgical and milling facilities for iron, copper, lead, silver, and gold to aid industrialization and finance; creating transportation and communications infrastructures, including railroads and a telegraph network, as well as a network of lighthouses; building shipyards from which would develop a navy and merchant marine corps; and conducting geological, geographical, and maritime surveys of Japan to facilitate in its development and defense.[23]

The Ministries of Finance (*Ōkurashō*) and Civil Affairs (*Minbushō*) also had roles in industrial modernization. Prior to 1870, the Ministry of Civil Affairs was charged with, among other things, the promotion of private industry. During the Bakumatsu era (1853–67), domains such as Satsuma, Saga, and Mito began to establish factories or metalworking sites which utilized Western machinery and organization.[24] As the Meiji government began to absorb or purchase former *bakufu* or domain-operated factories and mines, it was the Ministry of Civil Affairs that controlled these facilities. It continued to do so to a lesser extent after the founding of the public works ministry. Likewise, the Ministry of Home Affairs was charged with promoting a wide range of activities. Originally founded to suppress popular unrest, especially among disaffected samurai, the Ministry of Home Affairs was also concerned with industrial promotion and the development of a viable merchant marine. Prior to 1881, when the Ministry of Agriculture and Commerce (*Nōshōmushō*) was established, ministry officials focused a good deal of their attention on handicraft industries, such as lacquerware, wood and ivory objets d'art, bronzes, fine metalwork, and agricultural promotion in the areas of silk, cotton, and tea.[25] Because of overlapping jurisdiction, the two ministries were often at odds creating a policy–politician rivalry between Itō Hirobumi and

Ōkubo Toshimichi representing the Ministries of Public Works and Home Affairs, respectively.[26]

The Ministries of the Army and Navy also had a hand in early Meiji industrialization under *shokusan kōgyō*. The benefactors of Tokugawa imports of Western technology, under the new regime, the military absorbed and expanded arsenals, shipyards, machine and iron works. Most historians would agree that the military's program of industrial modernization was beneficial for the dissemination of Western technologies, aiding or creating 'spin-off' industries such as the machine tool industry which later aided in modernizing the textile industry.[27] As will be seen in subsequent chapters, competition for funds and program direction, however, tended to undermine Ministry of Public Works projects.

Given the enormity of the task of modernizing Japan's industrial and transportation infrastructures, not to mention its economy, it is no wonder that so many ministries had a hand in industrialization efforts. In order to attain relative industrial and economic parity with the West, modernization ultimately relied on the importation and absorption of foreign knowledge and artifacts. Under the leadership of the Ministry of Public Works, this took place in architecture, mining and metallurgy, railroads and telegraphy, civil engineering and construction projects, textiles, and in the creation of a national mint. Although reliance on foreign knowledge, whether in the form of foreign advisers or artifacts, made it possible to build a 'modern' nation, it was not possible to sustain technological or economic 'modernity' in this manner. As such, the Meiji government simultaneously expanded its activities into the social realm, forging an educational system that would continue to nurture national development.

The relationship between foreign knowledge, foreign advisers, and the Japanese who sought both is more complex than is often portrayed. Hiring foreign advisers was considered a necessary evil; a temporary measure designed to help Japan catch up with the West. This was true in all spheres including technology, the economy, military, society, and government. It is clear that the government was building on a precedent set by the *bakufu*. During the Bakumatsu era, the Tokugawa began to hire French advisers to help with their effort to import Western mining, military, and shipbuilding technologies. At the same time, they intended to establish a school at their shipyard to train Japanese in shipbuilding so they could become technologically independent.[28] Because the government at times intended to import complete technological systems, it is often stated that the Japanese uncritically followed foreign advice. Nothing could be further from the truth. Although the Meiji government paid dearly for its foreign knowledge, they were the 'customers' and chose to do as they pleased with their 'purchased' advice. Over the course of the Meiji era, approximately 3,000 foreign engineers and scientists alone came to Japan.[29] Because foreign expertise came with a hefty price tag, it was typical that contracts were not renewed. A good example of this is the case of Louis Bianchi, the government's adviser at Kamaishi Ironworks who will be discussed in greater detail in Chapter 4. With few exceptions, advisers on average were retained for five years.[30] From the beginning, the government's intention was to replace foreign advisers and foreign knowledge with indigenous knowledge; Japanese engineers who had acquired requisite knowledge and experience to fill technical positions. As foreign advisers came to Japan to help build the nation and educate Japan's future engineers, technicians, and scientists, ensuring the brevity of their own tenure, many Japanese went to Europe and the United States to study.[31] This was the basic method by which Japan gained technological and scientific independence. It worked well in some areas, less so in others. Japan's mixed record of success prior to approximately 1895 is demonstrative of the complex nature of technology transfer as a process.

Meiji era study travel had its precursor in the last years of the *bakufu* when a good number of Japanese traveled, legally or otherwise, to Europe and the United States on observation and study missions. Many of the Meiji oligarches were beneficiaries of overseas experiences and this helped to shape their attitudes toward the West in general and technical education in particular. The infamous Chōshū Five for example, Itō Hirobumi, Inoue Kaoru, Inoue Masaru, Yamao Yōzō, and Endō Kinsuke, violated the *bakufu*'s ban on overseas travel to study at University College London in 1863. Each went on to become an important member of the Meiji government. Initial efforts to establish a national system of compulsory education began in 1872. In addition to drawing on Japan's tradition of education and comparatively high literacy rates, it was envisioned, among other things, that this system would create a pool of talented individuals who would help create a 'modern' nation. From the dual yet complimentary perspectives of *shokusan kōgyō* and *fukoku kyōhei* the public works ministry's creation of the College of Engineering (*Kōbu daigakkō*) was essential to industrial development.

Founded in 1873, the College of Engineering had its origins in an earlier Meiji institution, the Engineering Bureau or *Kōgakuryo*. Established in 1871, the Engineering Bureau was split into two separate schools the following year, an engineering college and a technical preparatory school. In 1877, following its dissolution, the engineering college was formally named the College of Engineering. When the Ministry of Public Works was reorganized in 1885, the College of Engineering fell under Ministry of Education (*Monbushō*) jurisdiction. The following year it, along with two other institutions, were combined to become Tokyo Imperial University.[32]

The Meiji government recruited a young Scottish mechanical engineer named Henry Dyer to become the College of Engineering's principal and establish its curriculum. Overall responsibility for the College remained the responsibility of then Vice Minister of Public Works, Yamao Yōzō who, like Dyer, had studied at Anderson's College in Glasgow. The College employed many foreign instructors, most of whom were British. It should not be surprising then that the primary language of instruction was English, and that students were also required to write their theses in English.[33] The College boasted many professors who would soon become prominent in their fields. This list includes John Milne, the 'father of seismology,' Josiah Conder, the 'father of modern Japanese architecture,' and physicist William Edward Ayrton, whose Tokyo laboratory in applied electricity was the world's first.[34]

Engineers trained at the College of Engineering undertook in a six-year curriculum. The first two years were general and preparatory in nature. Students would spend their time in the classroom studying general science, mathematics, and engineering theory. The remaining four years were specialized and divided into two years of alternating theoretical studies and practical application, and two years that have been characterized as 'on-the-job training' where Dyer placed his students at various Ministry of Public Works project sites around the country.[35] Dyer's curriculum was revolutionary for its time and thus controversial. From the German and French perspectives the curriculum was deficient in engineering theory; British engineers criticized the curriculum for lacking enough practical training.[36] The latter point drew rather severe criticism from Richard Henry Brunton who argued that Japanese technical education was overly theoretical. He also stated that six years, only half of which were dedicated to practical experience, was too short a time in which to master the engineering arts.[37] Nonetheless, the College of Engineering trained young men in the fields of architecture, applied chemistry, civil engineering, electrical engineering (telegraphy), mechanical engineering, mining and metallurgy, and later naval architecture. Many of the College's graduates went on to become the men responsible for Japan's industrial 'take-off' during the latter half of the Meiji era. Many of the mechanical engineers responsible for industrial modernization of the cotton spinning industry such as Kikuchi Kyōzō were College of Engineering graduates.[38] Other graduates include Inokuchi Ariya, inventor of the Inokuchi centrifugal pump, chemist Takamine Jōkichi, who discovered adrenaline, and Tatsuno Kingo, one of several architects responsible for designing Victorian Japan.[39]

One of the greatest areas of change in Meiji Japan, at least in terms of the conspicuous and popular absorption of technological systems, was in transportation. Prior to the Meiji Restoration most Japanese got about on foot. A privileged few traveled by *kago*, a seat mounted on two long poles carried by porters, fewer still rode horseback. Replacing the *kago* in the 1870s, was the *jinrikisha*, or rickshaw as it is better known in the West. Most long distance transportation in Japan, especially for shipping goods, was done through coastal shipping networks. The greatest impact on land transportation, however, was the development of a railroad under the auspices Ministry of Public Works.[40] First introduced to Japan in late 1853 by Russian Admiral E. V. Putiatin in model form, and then again in March 1854 when Commodore Perry's men constructed a quarter-scale railroad as a gift intended to demonstrate America's industrial prowess, railroads and locomotives quickly became part of Japan's 'modern' technological landscape.

After being approached by Sir Harry Parkes, British Minister to Japan in late 1869, Meiji officials decided to build Japan's first railroad line from Tokyo to Kobe. They had been fending off proposals for foreign railway concessions and Parkes' suggestion came at an opportune moment. One point that helped the government decide in favor of railroad construction was the knowledge that a national railroad would be demonstrative of their authority. By linking distant provinces with the capital, the new government would achieve a kind of national unity unlike the *bakufu*. A railroad would also begin the transformation of Japan into a 'modern' nation.

Originally the line was to run through the center of the country, along the Nakasendō, but a combination of difficult terrain and expense moved the line to a coastal location, the Tōkaidō.[41] Construction began in April 1870 based on British colonial standard (narrow) gauge. The line, which went initially from Yokoyama to Shinagawa, a mere 15 miles (24 km), was in operation two years later. Several more miles of track were laid later in 1872, extending the line to Shinbashi, Tokyo.[42] While British engineers and Japanese workers were laying track in the Tokyo–Yokohama area, a second group of engineers were busy surveying for a line between Osaka and Kobe. Opened in May 1874, this line was extended to serve Kyoto in 1876 and Otsu by the end of the decade. The latter section, which included the Osakayama tunnel and several mountain passes, was designed and built by Japanese engineers who were graduates of the Engineer Training College.[43]

After several years of financial difficulty, spurred on by having to quell rebellions and the rising cost of importing foreign knowledge, the government agreed to the private construction and operation of railroad lines. Thus was born Nippon Railway in 1881.[44] By the end of the decade, there were five private railways in operation. Japan's railroads, national and privately owned, continued to expand at a fairly even pace despite a brief slowdown which accompanied an economic recession in the early 1890s. On the eve of railroad nationalization in 1906, there were 37 private railway companies with a total of 3,239 miles (5,213 km) of track. Government railroads accounted for 1,499 miles (2,413 km).[45]

In its early years, the Meiji government employed some 200 foreign engineers and technicians to help with railroad construction. By the end of the decade, however, Japanese engineers had replaced their foreign counterparts. This was made possible by what may be considered an extension of the government's policy of import substitution. Following what was to become a well established practice, British engineers trained Japanese, many of whom were College of Engineering graduates, in the latest Western railway technologies. Specifically addressing the technicalities of railway design, the government also established an Osaka-based Engineer Training College in 1877. Other future Japanese railway engineers studied abroad.[46] When they returned, they, along with their domestically trained equivalents, assumed technical and managerial positions in the Ministry of Public Works' Railway Bureau or the other ministries that had a hand in railway development.

Development of a railway system was not uncontroversial. Because of the immense capital requirements of railway construction, and knowing that private investors would hesitate to get on board, the Meiji government was forced to secure foreign loans. This move brought cries of official corruption and of selling out the country. Ōkuma Shigenobu and Itō Hirobumi, two leading proponents of railway development, also faced direct challenges from other ministries, including the military. Army officials considered railway expenditures frivolous when the country should use its limited funds to 'strengthen military affairs.'[47] The short-sightedness of the military is ironic. In 1877, less than a decade after delaying construction of the Tokyo–Yokohama line between Shinagawa and Shinbashi to reinforce its objections,[48] the army was a major beneficiary of the railroad upon which it relied to transport troops when putting down the Satsuma Rebellion. Other railway opponents argued that a network of roads would serve the nation just as well and at a fraction of the cost.[49]

Accompanying the development and spread of the railroad was the telegraph, of which a national network was largely complete by 1878. This had also been introduced to Japan during the Perry Mission, when Meiji leaders quickly recognized the importance of rapid national communications. Despite Tokugawa experiments with the technology, considered one of political importance, there was little success prior to the Meiji Restoration beyond a short line within Satsuma *daimyō* Shimazu Nariakira's castle compound.[50] By the following year, in 1869, Japan's first telegraph system was established. By 1870, it was available for public service between Tokyo and Yokohama. Based on British designs and imported equipment, and made possible through the assistance of British engineers and Japanese workers, Japan moved to expand reliable communications between Tokyo and Nagasaki. But the telegraph too, was not without its attackers, literally. During 1873 alone, disgruntled samurai drew their swords and destroyed hundreds of telegraph poles.[51]

As with the railroad, Meiji officials in general and especially those in the Ministry of Public Works understood the strategic and economic importance of the telegraph system. As a result, the Japanese government was sure to maintain direct control over both systems.[52] Although the telegraph initially relied on foreign knowledge, in the form of either British or American engineers and technological artifacts, within a decade foreign engineers had once again been replaced by their Japanese counterparts.[53]

As with so many other facets of Meiji industrial development, the modernization of mining, whether iron, non-ferrous metal, or coal, was undertaken with several objectives in mind. On the one hand, the government needed to secure the nation's natural resources to prevent their untethered exportation. On the other hand, the government sought to identify and regulate commodities for export. At the same time, Japan needed raw materials with which to build a viable military and to 'modernize.' The result was an overall program that sought to increase and regulate the supply of gold, silver, and copper for currency, to export copper and coal to foreign markets, to identify and exploit sources of copper, iron, and coal for the military, and to increase iron production to build a 'modern' nation.

The Meiji government inherited a mining industry that had been in decline for nearly a century and all but collapsed 40 years prior. Despite the availability of technologies that could have helped the industry, Tokugawa mining developed very little. To reverse this trend the government and private concerns hired scores of foreign advisers, nearly 100 for the non-ferrous metal mining industries alone. Men such as Francisque Coignet, J. G. H. Godfrey, Benjamin Smith Lyman, Edmund Naumann, and Curt Netto surveyed the condition of Japan's mining and metallurgical industries. Based on their observations, which were not always quite accurate, they made their recommendations.[54] Importantly, these men looked beyond technological deficiencies and also identified managerial problems.

Actual mining operations were inefficient. Regardless of what was being mined, tunnels were very narrow and meandered underground as they followed only the riches veins and only when digging was easy. Speaking of the Ikuno silver mine for example, Coignet described the tunnels as being only a few feet tall and wide, and winding in all manner with little observable direction.[55] All labor was done by hand; men attacked the ore-bearing rock with picks and chisels while women and children hauled out debris and ore in baskets. Once at the surface, only the best ore was kept. This was a rather wasteful practice because a good deal of metal could have been extracted from material that ended up as tailings. Water and ventilation were also problems in early mines. Although there were sophisticated pumps available, most mines were drained by buckets, simple bamboo hand pumps, or abandoned. Similarly, miners benefitted little from any advances in ventilation. At best, miners could expect to see a wheel-driven fan at the entrance to the mine. Refining fared little better. Ores were pulverized by hand using hammers, and metals were extracted by liquefaction, that is heating the crushed ore in a crucible until the metals melted and ran out.

If there was one point on which all the advisers agreed, it was that managerial practices at the mines were detrimental to efficient production. Japanese mining in the Tokugawa and early Meiji eras operated under a subcontracting system. At any given mine, the mine owner had little to no authority. Mines were run by 'bosses,' subcontractors who hired the miners, who made all decisions regarding where individual miners worked, set prices for ore, evaluated the ore, and sold it to the mine owner. Because there was more money to be made in areas with richer ore and easier digging, much valuable metal was left underground.[56] Miners and bosses also withheld large amounts of ore that could later be privately bartered. Practices such as these were common at most mines well into the late 1870s.[57]

Following the collapse of the Tokugawa Shogunate, the Meiji government either purchased or appropriated former *bakufu* or domain mines. Others were sold to private enterprises. Regardless, there was a significant push to improve Japan's mining and metallurgical capacities. Modernization of Japan's mines took a number of forms, only some of which were technological in nature. One of the earliest adopted innovations was the use of explosives, which was first introduced in 1867 by Erasmus Gower at the Sado mines and Raphael Pumpelly at the Yurup lead mine. Although a significant improvement, the use of explosives did not allow changes in mining practice to keep pace with refining.

Mechanization first came to mining as part of the refining process. This was done to maximize the amount of metal extracted from the raw ore. At the Ikuno mines for example, water-powered stampmills were imported to crush the ores prior to sorting and amalgamation. Similar methods were adopted at Sado's silver and gold mines. The adoption of the latest Western machinery for crushing the ore was a major improvement over the traditional method – hammers. Through amalgamation, the process by which gold and silver were extracted from ore by mixing with mercury, yields were also vastly improved. Large quantities of otherwise discarded metals were recovered. Netto noted, however, that because the mining sector was unable to keep pace with the improvements to refining, the results were not as spectacular as one would have hoped.[58]

Technological modernization of mining took place on many fronts. Because of the modernization of refining, it was profitable to extract metals from ores that had previously been ignored. Therefore, mine bosses and owners started to sink new shafts as galleries, that is a vertical shaft with horizontal branches. Other changes also worked to improve efficiency as well as safety, although the latter was not a concern. At Ashio copper mines, for example, some of the more important improvements included sinking deep vertical shafts, improved ventilation and drainage, replacing the miner's bamboo torches with oil lamps, and installing a tramway to move the ore from mine face to the refinery.[59] Similar innovations plus others such as imported steam-powered hoists for vertical lifting and machinery for cutting the rock face were seen at the government's mines.

As may be expected, many of the improvements to mining were a source of resistance. Possibly the greatest impediment to the modernization of Japanese mining was its managerial system. Mine bosses had a stake in keeping things the way they were and mine owners, whether private or the government, could not ignore this fact. At Ashio, owner Furukawa Ichibē side-stepped mine bosses by opening new shafts in areas that had not been subcontracted. He also purchased the rights to a rich strike from a mine boss, thus eliminating the middleman once again.[60] Eventually Furukawa began to replace the paternal relationship between miner and mine boss with corporate paternalism. Although not fully successful, it had positive effects.[61]

The government's solution to its managerial problems was to issue directives. As may be expected, this had little effect. In 1876, they tried another method, education and on-the-job training. Managers at the government-owned Ani copper mines sent 13 miners to Ikuno silver mines to be taught the latest Western mining practices with the hopes that they would adopt a more 'modern' attitude toward management.[62] This program, although maintained for some time and adopted by private mines as well, made little headway in breaking the centuries-old system. In the end, however, the government won. After nearly five years of maneuvering between the government and the mine bosses, the system finally came to an end in 1881.[63] Miners became contract wage workers and the former mine bosses took up supervisory and managerial roles.

As with other industries, Japanese began to receive technical training in mining and metallurgy almost from the start. In their reports to the Meiji government, Coignet, Netto, and Adolf Mezger all recommended that the government establish training centers at their respective mines. Coignet had a grand plan for a regional center based at Ikuno to train engineers that appears to have never materialized. In 1869, however, the government establish a school at which to train mining engineers; within three years, it had trained more than a dozen graduates. Eventually the Ministry of Public Works and the College of Engineering would continue to educate Japan's mining and metallurgical engineers.

The first Japanese coal mine to utilize Western techniques was the Saga domain mine at Takashima in what is now Nagasaki. Since the discovery of coal at the site more than a century and a half prior to its modernization, coal was extracted at Takashima from open pits. Miners would break away chunks of coal with picks and chisels until the seam ran out, the digging became too difficult, or the pit became too deep. In 1859, a Scottish merchant, Thomas Blake Glover, convinced *daimyō* Nabeshima Naomasa to modernize his mine.[64] After securing the financial backing of Jardine, Matheson & Co., Glover imported steam-powered drilling, lifting, hauling, pumping, and ventilation equipment to the site. He brought in British mining engineers and technicians to sink Japan's first Western-style (vertical) mine shaft in 1868. The following year, miners struck a major coal seam and Takashima became Japan's first modern colliery. Takashima retained that distinction until 1888 when the Miike coal mines underwent modernization.[65]

Takashima went through various incarnations as a private and government-owned mine, all of which relied on foreign knowledge either in the form of technological artifacts or British mining engineers. Over time, the British engineers were replaced with graduates from the College of Engineering. It was not unusual, however, to find foreign advisers working for private companies well into the twentieth century despite the existence of qualified indigenous engineers.

The remarkable part of the Takashima story, in someways greater than its technological development, is its capitalization. This mine was opened and operated almost exclusively with foreign funds. This was the case when Glover first approached Saga domain officials and continued to be the case even after the Meiji government passed its 1873 mining law that nationalized mineral resources and excluded foreign financial interests. The government assumed control of Takashima in either 1873 or 1874 and operated it for perhaps a year. In a clear violation of the new mining law, the mine was purchased by a soon to 'retire' government official, Gotō Shōjirō, at a bargain price with money borrowed from Jardine, Matheson & Co.[66] After years of financial wrangling, mismanagement, and legal battles, the mine was purchased by Iwasaki Yatarō, owner of the Mitsubishi Mail Steamship Company. It remained the leading producer of coal in Japan until 1890.[67]

The facet of Meiji industrialization with which most people are familiar is the government's so-called model factories. By and large, historians and economists consider the government's efforts in this regard to be either technological or economic failures, or both. What follows is a brief examination of government and private initiatives to modernize the cotton spinning industry. As will be seen, consensus falls on the government's failure and private enterprise's success. Regardless, the thread which is woven throughout the story is the importance of absorbed foreign knowledge for Meiji industry and the difficulties associated with technology transfer.

Following visits to the United States and Europe prior to the collapse of the Tokugawa *bakufu*, many Japanese who would become influential in the new Meiji government became well aware of the state of Western industry and the importance of industrial modernization for Japan. Ōkubo Toshimichi, a major proponent of industrialization who became finance minister in 1871 and home minister in 1873, came away from an 1869 visit to Satsuma's Kagoshima Cotton Spinning Mill convinced of the necessity of Japan's industrial modernization. This view was only compounded and confirmed following his participation in the Iwakura Mission. Ōkubo was both impressed with, or perhaps even in awe of, the state of Western industry and society. Yet he was confident that Japan too could be a 'modern' industrial nation. Based on these experiences Ōkubo advocated a policy of *fukokuron* (rich nation thesis or thesis on national prosperity), which supported industrialization over militarization.[68] As may be expected, Ōkubo's position added to inter-ministerial competition.

Along with Ōkuma Shigenobu, Ōkubo believed that government initiatives in industrial modernization would pave the way for private enterprise. Ōkubo may have not originally believed that it was the state's duty to establish industrial enterprises, but he soon came to support that position based on the recognition that the capitally intensive nature of industrialization was beyond the means of much of the private sector.[69] As a result, and through the Ministries of Public Works, Finance, Civil Affairs, and later Home Affairs, the Meiji government established a number of industrial enterprises for the purpose of industrial promotion and import substitution. Building on the mines, mills, and shipyards acquired from the *bakufu* and former *daimyō* domains, the Meiji government opened facilities such as Tomioka Silk Filature and Kamaishi Ironworks discussed in Chapters 3 and 4 respectively, Akabane Machine Works, Aiichi and Jukki Spinning Mills, Shinagawa Glass Factory,[70] Fukagawa Brick Factory, and the Senjū Woolen Mill.

Industrial modernization of cotton spinning is a good example of the dual nature of Meiji industrialization. To this point we have been looking at initiatives that were primarily the concern of the state. And while modernizing the spinning industry was a goal of the Meiji government, its role was limited, with much of the work being done by individuals or through local initiative. As previously mentioned, the Meiji government absorbed Satsuma domain's spinning mills in Kagoshima and Sakai. The original mills were proto-industrial in nature. Shimazu Nariakira assembled a number of traditional spinners and weavers under one roof and later constructed a water-powered weaving shed. The machinery was wooden and locally made. Although Shimazu contemplated importing foreign-made looms, they were cost prohibitive. He also realized that Western looms would not work with traditionally spun thread and feared that complete industrial modernization based on Western technology would wreak economic havoc with farmers who depended on wages from spinning by-employment.[71] By the end of the 1850s, however, it was clear that Satsuma's mills, relying on traditional technologies, could not compete with the Western imported cottons entering the country through the various treaty ports.

Following Shimazu's death, a group of domain officials and students were sent to the United States and Europe on a study mission. The goal of this highly illegal mission was to study and acquire modern Western weaponry, industrial technology, and perhaps conclude agreements of support between Satsuma, Britain, and France.[72] Several members of the group, including Niiro Gyōbu and Godai Tomoatsu, were charged with studying the British cotton industry, purchasing modern spinning machinery, and hiring British engineers to oversee the founding of a new Western-style spinning mill.[73] Because neither man had the technical expertise on which to base his decision, the pair approached the Lancashire firm of Platt Brothers with the hopes of getting a 'package deal' of sorts. In early 1866, Niiro and Godai concluded their arrangements with Platt Brothers. One year later the spinning frames, with a total of 3,648 spindles, arrived in Nagasaki.[74] By June 1867, all the machinery was installed and the factory, small by Western standards, began production.

Initially Kagoshima ran under the stewardship of the British engineer Edward Holme, who was sent to Japan by Platt Brothers. Within a year of opening its doors, however, Holme and the other British engineers had returned to Lancashire, leaving the mill in the hands of local managers who lacked the expertise to efficiently run the facility. An additional problem for the mill was its location far from the center of the cotton trade. Domain officials attempted to remedy the location issue with the establishment of the Sakai mill in 1871. With 4,000 spindles, Sakai Spinning Mill began operations. The mill benefitted from Kagoshima's technical experience but was, nonetheless, not commercially viable. In May 1872, the Ministry of Finance purchased the mill.

Fearing the effects of growing dependence on imported cotton goods, the Meiji government decided to actively support the industry. It proposed to open two mills, one in Aichi prefecture and a second in Hiroshima. Officials also decided to import Western machinery to sell directly to private firms at no interest. This began what has been called the 'period of 2,000-spindle cotton spinning plants' during which time 14 spinning mills each with 2,000 spindles were established.[75] The original group of ten firms, Jukki Spinning Mills, were opened as government factories in 1880 with British-made equipment. Nine of these were later sold.[76] Like Satsuma's efforts to modernize its cotton spinning industry, the government's mills were also plagued by a lack of skilled engineers and laborers. Each of the mills had modern spinning equipment and the factories were often designed and set up by the same British firms which sold the machinery. Yet there existed significant barriers to the successful transfer of technology. One problem was a lack of practical experience in working with complex mechanical technologies. Another often-identified problem was the inability of British engineers to explain the intricacies of a complicated process to Japanese technicians who had a limited command of the English language. With only 2,000 spindles and under-capitalized, these firms were largely commercial failures which were unable to achieve the government's goal of cotton textile independence.

Successful modernization of the cotton spinning industry came to Meiji Japan in 1882 with a private-sector initiative led by Shibusawa Eiichi's Osaka Cotton Spinning Company. Shibusawa's project was significantly different from earlier initiatives. Although relying on imported machinery, it was thoroughly imbued with technical expertise. Because the Osaka Cotton Spinning Company relied on a combination of foreign and indigenous knowledge, it is possible to view it as a microcosm of the Meiji industrial education experience. In the beginning, the company relied on foreign engineers who maintained an advisory relationship with the firm for quite some time.[77] From its inception, however, Osaka Cotton Spinning Company, or perhaps more accurately Shibusawa Eiichi, was determined to rely on indigenous knowledge.

Prior to starting construction on the mill, Shibusawa approached Yamanobe Takeo, who would become the company's first engineer and manager. Yamanobe, who had been studying insurance at University College London, was recommended to Shibusawa by a mutual acquaintance, Tsuda Tsukane. After some consideration, Yamanobe abandoned his present course of study and transferred to King's College where he would study mechanical engineering. He would later travel to Manchester and Lancashire, eventually working at the Rose Hill Mill in Blackburn.[78]

Yamanobe would later be responsible for many of the key decisions related to the design of the spinning mill. He selected the site, chose to rely on steam power rather than water power, as had earlier been planned, selected the original mule spinning-frame technology and made the later decision to change the mill over to ring-frame technology.[79] Yamanobe is also credited with the 1896 decision of abandoning the use of kerosene lamps, as was common at the time, in favor of electric lighting. This was a particularly significant use of a new technology to reduce the threat of fire given that the mill was in operation 24-hours a day.[80]

The Osaka Spinning Mill also differed in its financial backing and scale of operations. It is possible to state that one contributed to the other. Shibusawa raised 280,000 yen in investor capital from merchants and former *daimyō* from Tokyo and Osaka.[81] This kind of financial support enabled Shibusawa to build the country's largest cotton mill in terms of its physical space, nearly 6 acres (24,282 sq. metres), its 10,500 spindles, which made it more than five times larger that its nearest competitor, and to follow Yamanobe's recommendation to use steam power.[82] Another innovation, round-the-clock operation, helped the mill achieve economies of scale.[83] In slightly less than two years of operation, Osaka Spinning Mill began to report profits and paid an 18 percent dividend to its investors.[84]

Yamanobe was more than Osaka Spinning Mill's engineer and manager: he represents the mid-Meiji move toward technological self-sufficiency. Following Yamanobe were other Japanese engineers who became managers of major cotton spinning mills. Kikuchi Kyōzō, Saitō Tsunezō, Hattori Shun'ichi, and Takatsuji Narazo for example all trained in mechanical engineering at the College of Engineering. Following graduation some took positions in government enterprises such as the National Mint or Yokosuka Naval Yard. All eventually traveled to Britain where they would gain practical experience in the Lancashire spinning industry.[85] Upon their return to Japan each would find positions in private companies such as the Mie or Settsu Cotton Spinning Mills, where they would begin to transfer their foreign and indigenous knowledge. In part, men such as these made possible the rapid development of, and investment in, post-1886 Meiji industry. Although Shibusawa Eiichi's business acumen cannot be denied, the successful modernization of the cotton spinning industry rested on accumulated technical knowledge and its practical application in the hands of Japanese engineers and managers.[86]

It may appear that at times traditional industry or attitudes were more of a hindrance than a contribution to modernization. This was not the case. In general, the role of traditional industry in Meiji industrial modernization was most pronounced in private enterprise. This was especially so in textiles, machine tools, and small-scale mining, but it should also be noted that the Japanese technicians who worked for the railroad or workers who built the government's factories were traditional craftsmen. For traditional industries such as silk reeling, cotton spinning, or weaving to compete with imports and large-scale factories, they needed to modernize and respond to market conditions. As will be seen in the next chapter, industrial modernization in silk reeling by following the government's example was beyond the economic means of most entrepreneurs. As a result, and whether moved by personal ambition, a declining local economy, or an ideology of progress, entrepreneurs turned to local artisans to help them 'modernize.'

The case of Gaun Tokimune's *garabō* or rattling spindle is a well-known example of an indigenous innovation that served small- and medium-sized enterprises in the textile industry. Gaun's original invention was a simple, wooden spinning machine where raw cotton was packed into bamboo tubes and then drawn out and twisted into a coarse yarn through a series of rollers. Although it was a labor-saving device, it was appreciated by neither his family nor local producers.[87] Gaun worked to improve his machine in the 1870s as Japan's cotton industry began to feel the threat of foreign imports. This time, cotton spinners embraced the *garabō*. They copied and improved it further. The number of spindles, for example, was increased and the hand crank was replaced by water-power. By the late 1880s, *garabō* were the mainstay in hundreds of small- and medium-sized textile factories.

Throughout the latter half of the Meiji era, Western power looms were imported to Japan at the behest of large textile manufacturers. These looms, however, were unsuited to the production of the narrower cloth preferred by the domestic market.[88] Western-made looms were also beyond the financial means of most small or medium enterprises. Looking to remedy this problem was Sakichi Toyoda, who began developing wooden hand looms in 1887.[89] In 1891, he patented his first hand loom that would increase weaving efficiency by 50 percent. Unfortunately Toyoda's loom was not a success because of a French invention introduced at the same time that was more cost effective. Undaunted, Toyoda continued to experiment with different machines that would improve the efficiency of Japan's cotton textile industry. He invented a successful yarn-reeling machine and, in 1896, produced a successful narrow wooden power loom. Toyoda's loom increased worker efficiency, reduced costs, and improved quality. Toyoda continued to invent and improve his looms. Eventually they were made of iron, which allowed for the use of fully interchangeable parts. The culmination of his efforts was the creation of the world's first fully automatic loom in 1924.

In Japan's traditional weaving districts, other craftsmen worked to invent power looms that would fill the voids left by Toyoda, such as looms for weaving stripped cloth. Silk weavers similarly sought ways to increase production of their primary export item, *habutae*, a plain silk fabric. The Yamagata prefecture town of Tsuruoka benefitted from the presence of two silk loom inventors, both of whom had the financial means with which to purchase machine tools. The first power loom for weaving silk was invented in 1898 by Saito Toichi, a large landowner. With the technical help of skilled craftsmen from Tokyo, Hirata Yonekichi, a major *habutae* trader, invented another model in 1907.[90]

There were similar advances in the silk reeling industry. In areas of the country that were the traditional hubs of silk reeling activity, small- and medium-sized enterprises made a variety of adaptations to their reeling practices – all of which were based on the efforts of local craftsmen. In the early 1870s, craftsmen fashioned their own versions of an Italian-style *tavelle*, a *croisure* or crossing device, and attached them to traditional Japanese reeling machines.[91] Others added these devices

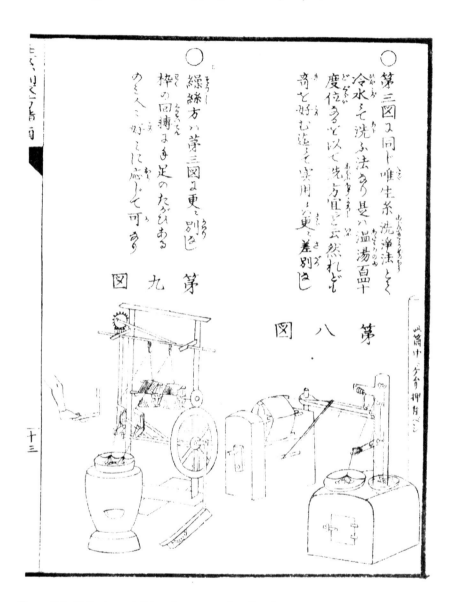

Figure 2.5 At the lower left is a foot-powered reeling frame utilizing Italian-style *tavelle croisure* devices (*Source*: Tachi, 1874: 13).

Figure 2.6 (*Right*) Ganged traditional reeling machines powered by a waterwheel that use Italian-style *tavelle croisure* devices (*Source*: Tachi, 1874: 16)

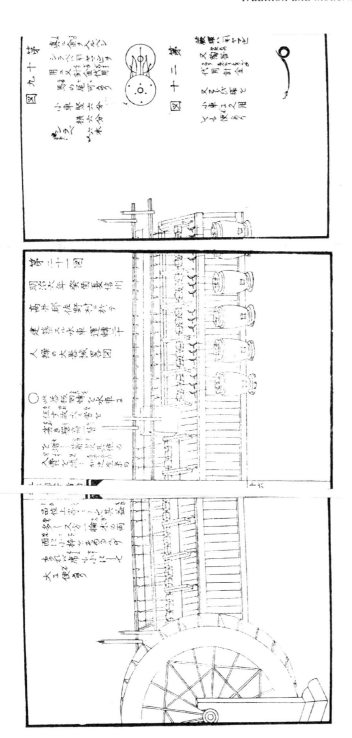

and used water-power to drive 20 or more traditional reeling machines which had been joined together.[92] More major changes include completely building Western-style reeling machines or hybrid Japanese–Western machines from wood. The use of steam and steam boilers in the reeling process is a related innovation derived from the growth of a domestic machine industry.

Shortly after the Meiji government opened Tomioka filature in 1872, filatures in Gunma and Nagano prefectures sought to adopt the steam reeling methods seen at the so-called model factory. Because importing cast iron boilers was cost prohibitive, they turned to local coppersmiths and metal casters. Relying on traditional techniques, these craftsmen built boilers suitable to the task.[93] By the end of the decade these Japanese-type boilers were used in more than 150 Nagano prefecture filatures.[94] Building on the success of these early boilers, Maruyama Yasaburō, a Nagano prefecture iron and brass goods manufacturer, invented a thin-plate boiler based on the design of a Cornish boiler. Although having its limitations, Maruyama's boiler was more efficient than a traditional Cornish boiler and, because it used less material in its construction, was less expensive to produce. Throughout the 1880s and 1890s, Maruyama continued to improve his boiler. Eventually his boilers were able to withstand the pressure required for use with steam engines, thus enabling the growth of Nagano's silk reeling industry.[95]

Throughout the Meiji era, Japan's program of industrial modernization moved forward through the combined efforts of the state and private enterprise. Although much attention has been given to Japan's industrial revolution in the decades after 1885, by introducing early industrial modernization initiatives, I hope to have shed light on the importance of the early years as well. *Shokusan kōgyō*, if it can be considered a formal policy at all, was haphazard at best. This is not meant as criticism. The Meiji government was under constant pressure from within and without. As such, various ministers and ministries, with the good of the nation in mind, competed to put forth their favored projects with limited funds. Itō Hirobumi and Yamao Yōzō at the Ministry of Public Works were proponents of heavy industry and engineering education. Ōkubo Toshimichi, the Minister of Home Affairs, also supported education but was an advocate of agriculturally based projects. As we shall see in the following chapters, the Meiji government's understanding of industrial 'modernization' and the means by which to 'increase production and promote industry' would continue to evolve even as two of its major projects were underway.

If there was a single agreed-upon element of the modernization process, it was that Japan needed to be technologically independent. While it was necessary to rely initially on foreign knowledge, the plan had always been for it to be replaced with a domestic alternative. Many of the *yatoi* (foreign employees) themselves, whether implicitly or explicitly, supported this policy. Importantly, the basis of Japanese modernization and subsequent industrial revolution can be said to rest firmly on the education provided at the College of Engineering, the many workshops under the supervision of the Meiji government, through overseas study, or at private enterprises which employed foreign advisers. Traditional knowledge and skills also played an important role in the modernization process. As we shall see, these contributions would often be ignored by a 'modernizing' government.

3 Iron machines and brick buildings
The material culture of silk reeling

On a spring morning in 1871, a group of workmen gathered in a remote valley in Gunma prefecture to break ground on what was to become the Meiji government's model filature. According to public pronouncements, Tomioka Silk Filature was being built as part of a government initiative to improve the quality of Japan's raw silk through the dissemination of Western silk reeling technology.[1] As a model factory based on Western methods, Tomioka was supposed to be a technology showcase and a training center. The factory, an impressive complex of brick buildings, housed the latest French-made machinery powered by steam engines. Visitors, including many government officials and the royal family, flocked to the backwoods of Gunma to marvel at Japan's technological progress.

The actions and comments of Tomioka's visitors, whether entrepreneur with visions of improving business, future trainee, or self-congratulatory bureaucrat, are indicative of the factory's greatest value and purpose, one not discussed at its inception. Based on the government's behavior, recommendations in state-sponsored silk reeler's manuals, and the comments and actions of local reelers and the women trainees, it becomes clear that the true value of Tomioka Silk Filature was its symbolism. As a model on which to base the transfer of silk reeling technology to local producers, as a site at which to train women in practical reeling methods, and as a model business venture Tomioka was, at best, only partially successful. As a model of 'progress and civilization,' as an element representative of a society trying to advance under a Victorian value system, however, Tomioka was ideal. Every component of Tomioka was symbolic. Brick, cast iron, steam engines, and coal were synonymous with 'progress and civilization.' These materials, regardless of practical value, were described in terms of their strength, permanence, and precision. Whether anyone could actually replicate what they had seen at Tomioka never seems to have been the government's concern.

Simply labeling the Tomioka vision 'modernity' or 'imported civilization' betrays the complexity of the situation. On the one hand, Tomioka was symbolic of the new Meiji government's need to situate itself politically above the semi-autonomous domains; on the other, it exemplified the government's relationships with Britain and France. Silk, particularly Tomioka silk, also became the yardstick by which the Meiji government measured itself against other European and Asian nations. Much as local reelers advertised their raw silk, reeled by traditional methods, as machine-made, or as entrepreneurs made exaggerated claims that their filatures were based on Tomioka's technology, the Meiji government packaged its industrialization efforts under what it perceived to be the most efficacious labels.[2]

Mechanization of the silk industry: initial contact

In late 1869, Vice Minister of Finance Itō Hirobumi was approached by Freiderich Geisenheimer, the manager of a French trading company in Yokohama, who complained about the declining quality of Japanese silk.[3] As a remedy, Geisenheimer recommended establishing a model filature where Japanese silk producers could learn Western reeling techniques to the commercial (and technological) benefit of all concerned. More than changing the face of Japan's domestic reeling industry, the merchant was concerned with profit. Itō demurred, citing possible treaty violations as the reason. Insistently, however, Geisenheimer suggested a Japanese-run facility that relied on French capital and technology imports, but once again Itō declined. He was intrigued, however, by Geisenheimer's insistence as to the profitability of a model factory based on Western technology. Shortly thereafter, Itō initiated discussions between the Ministry of Civil Affairs and the Ministry of Finance regarding the feasibility of such a project. After some deliberations, officials within the two ministries decided to hire foreign advisers and proceed with the project. Itō contacted Shibusawa Eiichi, head of the finance ministry's Taxation Bureau and the only government official with any silk-related experience, to help find a suitable Western adviser.[4] In February 1870, Itō and Shibusawa went to Tsukiji, Tokyo where they approached Lieutenant Albert Charles Dubousquet to help the government find a silk reeling expert. On the recommendation of Ōkuma Shigenobu, they also went to Yokohama where they again spoke with Geisenheimer about the venture. Geisenheimer, the man who initiated the entire process, was the branch manager of the Lyons-based silk wholesaler Hècht, Lilienthal and Co. He and Dubousquet recommended that the government hire Paul Brunat, a young Frenchman who had been working for Hècht, Lilienthal and Co. in Yokohama for about two years.[5]

Brunat had been sent to Yokohama by the Lyons silk wholesaler, where he had worked in the silk industry since he was a teenager.[6] The four men visited Brunat in June 1870, at which point Itō and Shibusawa decided to hire him as the government's primary foreign silk reeling adviser. Shortly thereafter, Brunat was given a provisional contract. His final contract was signed in November of the same year.[7]

Profit and a steady supply of high-quality raw silk seem to have been the prime motivating factors in Geisenheimer's and Dubousquet's recommendation of Brunat who, for the first three years of his term of employment with the Meiji government, was simultaneously employed by Hècht, Lilienthal and Co.[8] Learning of the new partnership, the French community in Yokohama reacted with great enthusiasm and the natural assumption that the Tomioka–Yokohama–Lyons connection would translate into a return of high-quality raw silk imports, greater profits, and equipment purchases from France.[9] According to the terms of Brunat's final contract, where Geisenheimer was one of the signatories, all the equipment for Tomioka was imported through Hècht, Lilienthal and Co.'s Lyons office.[10] If the history of the

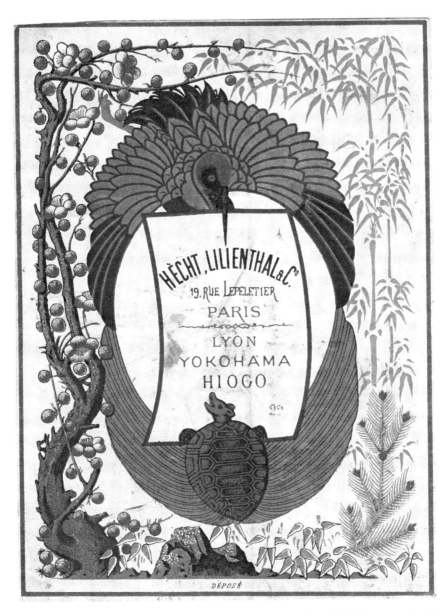

Figure 3.1 Hècht, Lilienthal & Co. raw silk packaging label, c. 1868 (*Source*: Author's collection)

Table 3.1 Raw silk exports from Japan, 1861–95 (annual average in kilograms)

Year	France	Britain	USA
1861–65	84,853	346,187	2,419
1866–70	229,703	364,029	21,107
1871–75	285,768	315,463	12,640
1876–80	470,957	320,241	176,238
1881–85	727,393	172,065	587,502
1886–90	772,692	84,006	1,110,715
1891–95	1,174,884	50,500	1,744,424

Source: Adapted from Sugiyama, 1988: 80

trade was a valid indication, the economic benefit of such a relationship was significant (see Table 3.1). As early as 1861, the export silk trade with France totaled some 2,600,000 francs. During the mid-1860s the value of the trade increased by more than a factor of seven, to 20,220,000 francs; in 1868, it again more than doubled.[11]

Perhaps hoping for a return to the late-Tokugawa days of greater French influence in Japanese affairs, some members of the French community in Yokohama apparently saw the new silk alliance as a way to limit efforts by Sir Harry Parkes to turn Japan into a veritable British colony.[12] Largely the result of Leon Roches' efforts and citing the Société Générale Française d'Exportation et d'Importation (a government-arranged private trading company), French involvement in proposing and constructing the Yokosuka dockyard, ironworks, and arsenal, and French efforts to restructure and revitalize the shogun's army, historian Richard Sims characterizes France and Japan as having a 'special relationship.'[13] I would argue that this relationship was deeper than even Sims suggests.

As early as 1865, Sigismond Lilienthal, partner in Hècht, Lilienthal and Co., was instrumental in orchestrating arms-for-silk deals with the Tokugawa government. Complaining of unfair trade advantages, a possible silk monopoly, and hinting at the impropriety of trading silk for guns, British chargé d'affaires Charles Winchester stated in a confidential memorandum from March of that year that Hècht, Lilienthal and Co. was accumulating quantities of raw silk 'for which no payment had been made and the value of which was far beyond the commercial resources of that firm.'[14] Two years later, in 1867, Alex von Siebold, translator and interpreter to the British legation in Yokohama, confirmed Winchester's suspicions noting in a private communiqué to Sir Harry Parkes that the *bakufu* had used raw silk to pay for the arms it had obtained from the French government.[15]

According to von Siebold's report, Roches' intention was not simply better trade relations: he was attempting a commercial monopoly. Through Roches' actions, the Société Générale Française d'Exportation et d'Importation, a theoretically 'private' trading company, became the shogun's agent in Europe providing the *bakufu* with contracts for 'guns and munitions of war' valued in excess of seven million francs, and offers to raise 'a loan of one million pounds sterling . . . in Europe

. . . in the name of the Japanese government.'[16] French and Japanese intentions were for the *bakufu* to repay its debts with produce, that is raw silk and tea.

Roches' dealings did not stop with commerce. His efforts on behalf of the *bakufu* seem to have been designed to 'support the present Tycoon [shogun], so as to make him powerful enough to put down all opposition even by force of arms.'[17] In Paris, the French government followed this policy, giving Satsuma's envoys to the 1867 International Exhibition a rather cool reception. Only after the Comte de Montblanc intervened on their behalf was Satsuma granted its own exhibition space, to be identified as separate from 'the Exhibition of the Government of the Tycoon' though still united under the Japanese flag.[18]

Just how far Leon Roches was willing to go to ensure France's and his own position is unclear. Although von Siebold doubted any involvement, the air of impropriety hung on some of his dealings with the *bakufu*. All *bakufu* financial transactions in Europe were managed by the Société Générale's banker Flury Hérand. Hérand was Consul-General to Japan in Paris, and he was also Roches' personal banker.[19]

Luckily for the British, the majority of Société Générale maneuvers failed as it was unable to raise the required capital. And although there appeared to be no end in sight to the arms-for-silk deals, with shipments of 40,000 rifles to Japan in late 1867, British fears of a *bakufu*-based French trade monopoly must have been assuaged with the 3 January 1868 coup d'état and subsequent events that marked the beginning of what was to become the Meiji Restoration. Writing from Paris on 28 January 1868, von Siebold noted that political changes in Japan meant that Roches' maneuvering was probably for naught.[20] France begrudgingly accepted the new reality, only moving to support the new Meiji government when it was clear that the *bakufu* was doomed. Upon receiving an invitation to an audience with the Meiji Emperor, Roches accepted not so much in recognition of his authority, but because he did not want to let Sir Harry have the honor alone![21]

Arms-for-silk deals continued after the fall of the *bakufu* and it appears that Hècht, Lilienthal and Co. maintained its presence in the illicit trade.[22] Geisenheimer's advances and recommendations may have been part of a French attempt to regain lost ground. Still fearing a Japanese government-led silk monopoly that would have adverse affects on the London trade, British officials and merchants in Yokohama continued to protest the practice, although the transactions were now of spurious value. Whether it was for trade or sale, the Meiji government was still left with the monumental task of improving the rapidly declining quality of Japan's raw silk. Market values were reaching all-time lows, the government's economic situation was no better; and although a foreign adviser had been hired, the technology upon which to base the mechanization of Japan's silk reeling industry was still undecided – or so it seemed.

Choice of technique: recommendation and rejection

From its inception, key players within the Ministries of Finance and Civil Affairs had different ideas about Tomioka Silk Filature. For some, it was to be a relatively humble facility, for others something more grand. Compounding, or perhaps reflecting this disparity is the Meiji government's lack of a clear vision of purpose for Tomioka.[23] To Odaka Atsutada, Shibusawa Eiichi's cousin, brother-in-law, and ultimately the facility's day-to-day supervisor, the facility was a business venture.[24] To others, Tomioka was a model facility that would serve to train Japanese silk reelers in modern Western methods. More importantly, however, for men like Shibusawa Eiichi and Sugiura Yuzuru, by utilizing the artifacts and ideologies of nineteenth-century Europe, Tomioka would serve as an exemplar of Western civilization in Japan.[25]

Paul Brunat recommended that the Meiji government build a sizable facility that utilized a Japanese–European hybrid reeling technology. Because the Tomioka area was already famous for its high quality *zaguri* reeling techniques, and because he did not want to disturb local production patterns, Brunat suggested surveying local reeling methods and incorporating them into Western mechanized reeling processes. He recommended locally produced machines, made of materials that were available in Japan, that is wood, and also suggested a local survey so as to ascertain what changes to existing Japanese reeling methods would be most beneficial. He also sought to find out how resistant local silk reelers would be toward changes in their industry. From the perspective of choice of technique, Brunat's proposal was relatively conservative. Although he recommended the incorporation of steam reeling techniques, steam engines to power the mill, and proposed a large facility – 300 basins – much of his plan was based on the use of local methods and technology.[26]

Government reactions, while seemingly enthusiastic, must have been a source of frustration for Brunat. On 29 November 1870, the day he received his final contract, he was ordered to return to France to purchase everything required, literally to import a fully operational French filature. Brunat was ordered to buy the latest cast iron reeling frames, steam engines, machine tools, and other equipment.[27] The directive called for a facility, the design of which was to be based primarily on the government's specifications.[28] Meiji leaders' desire to use the latest Western technologies bordered on obsession. The debate over fuel for the steam engines and boilers illustrates this especially well: Brunat recommended wood over coal because of the latter's exorbitant price. Meiji bureaucrats, however, opted for coal and not just any coal. It too, was to be imported from France![29] In the end, Tomioka would be based strictly on orthodox French reeling technology.

From what evidence remains, it appears that Brunat's proposal was rejected by a few men with little or no experience in the silk industry. Of the five officials directly responsible for Tomioka, one, Shibusawa Eiichi, had experience in sericulture. Odaka Atsutada was from a merchant family that dealt in indigo. The remaining three officials, Tamanoe Seiri, Nakamura Michita, and Sugiura Yuzuru, were former samurai with little or no commercial or technical experience. Two additional men, Ōki Takatō and Yoshii Tomozane, the officials of record in the Ministry of Civil Affairs, were also former samurai with apparently no technical or commercial background.

As discussed previously, many Meiji bureaucrats studied overseas, if only briefly, and this was an asset in their development of an industrialization policy. Shibusawa and Sugiura similarly studied in France in 1867, but how much they actually learned is questionable. Shibusawa himself admitted only to being capable of the most rudimentary French, making tasks such as shopping possible without an interpreter.[30] At issue, however, should not be whether these men had technical backgrounds, but that they chose to ignore the advice for which they were paying in the form of foreign advisers. For Meiji policymakers committed to industrialization, overseas experience legitimized cultural materiality as represented in *bunmei kaika* ideology. In sum, the government's actions with respect to technique/technology clearly indicate an ideological alternative agenda.

Choice of technique: economic perspectives

In an effort to better understand the Meiji government's position regarding the technology it chose to import and promote, it is appropriate first to briefly examine factors which are often considered important for choice of technique. The most basic arguments focus on economic factors which are considered determinative for technological choice. Proponents of this position argue that entrepreneurs will adopt the methods that give them the greatest output for the smallest input. Issues which add to the complexity of this model are the structure of related technologies within pre-existing industries, and the extent of the distribution of complementary or alternative existing technologies. Given these conditions, the 'proper' choice of technique for the Meiji government would have been the one which: (a) provided the most favorable ratio of increased productivity to minimal capital outlay; (b) fit largely within the structure of, and enhanced, the existing silk industry; and (c) could be adopted within the context of the existing technology. That is, adoption of a new reeling technology would not require completely abandoning the previous manufacturing system. As Penelope Francks notes in her study of innovation in Japanese agriculture, however, the aforementioned model does not take local conditions, such as the physical environment and climate, infrastructure, and availability of materials, into account; neither does it address such external issues as economic assistance or education dependency.[31]

From this perspective, the Meiji government's adoption of French reeling technology in the form of a turnkey facility should be considered economically irrational. Brunat's original proposal focused on building a filature that incorporated modern techniques, such as steam reeling and steam mechanization, into a context that was technologically and economically appropriate for conditions in Japan. He tried to address the issue of incorporating Western technology into existing methods. He urged that there be no break with tradition and suggested that the government incorporate recent European technological advances that would help raise the quality of Japanese silk and efficiency of the industry in general.

While economic issues figured strongly in Brunat's prospectus, most of his recommendations were rejected. He noted that importing all the necessary machinery was expensive, and repairing cast iron machines would be nearly impossible given the state of Japan's iron industry. The delays involved with obtaining replacement parts from France would be extreme and the costs exorbitant. The French process was also significantly different from local methods. It was more complicated and it would take a long time for workers, whether experienced or inexperienced, to gain any degree of proficiency. Brunat argued against any abrupt changes to the industry that would be inefficient or expensive. Where changes to the industry were necessary, Brunat's proposals were modest. Although the government, at least intermittently, appears also to have viewed Tomioka as a business venture, officials frequently ignored advice that would, at the very least, have reduced initial capital expenditures and increased productivity.

From a technical perspective Tomioka was irrational: its technology was too expensive and overly complicated. Further, from an economic perspective, in addition to the high initial capital investments required, the proposed filature was beyond the existing local market mechanisms in the 1870s. Simply put, the facility required too much in the way of raw materials. Local cocoon suppliers were unable to provide Tomioka with enough cocoons for it to run at full capacity, thereby adding to its inefficiency. Partly because the filature was located in a very remote area, still difficult to access to this day, the extra costs involved with importing raw materials from other parts of Japan created further problems: Tomioka's managers would not have had the financial resources with which to hire the additional workers necessary to bring the plant up to full capacity (assuming it had the raw materials).[32] It is clear that external conditions such as availability of raw materials, geographic location, and labor did not figure heavily in the government's plan.[33] As would be expected, with the exception of a few periods of severe economic entrenchment, the facility rarely turned a profit during its first decade of operation.

Technological alternatives and the materiality of 'civilization'

The chief critic of the government's plan was Hayami Kenzō, the man responsible for first importing Western mechanized reeling technology to Japan, and often considered the father of Japan's modern silk reeling industry. Hayami questioned the feasibility of the facility noting that its (French) technology was too expensive and complicated 'to be of any immediate benefit to Japan's silk reeling industry' or to the Meiji government. Based on his assessment of the government's proposal, Hayami also questioned whether or not Brunat was qualified. This was the first time anyone seriously questioned the government's plans,[34] and it would not be the last time that Hayami leveled an attack on government-led silk mechanization initiatives. It would also not be the last time the Meiji government chose to ignore the advice of its own advisers.

Hayami was in a unique position from which to criticize. Working originally for the Maebashi domain, he first established a retail silk outlet in Yokohama in March 1869. For a samurai, Hayami was a rather astute businessman. After meeting with the Swiss consul about the poor quality of Japanese silk, he surveyed Western merchants in Yokohama and confirmed statements that the primary problem with Japanese silk was not the raw material, as had previously been assumed, but the poor quality of Japanese reeling techniques.[35] Hayami was quite correct in identifying the problem. The Yokohama Chamber of Commerce, desperate for an end to the declining quality of Japanese raw silk had, in a rather unusual move, goaded the British legation into action. In the summer of 1869, F. O. Adams, Secretary to the British Legation in Yokohama, accompanied by silk inspectors from three Yokohama firms, including Paul Brunat, toured the silk producing districts in an attempt to identify and remedy the quality issue. Adams made two additional tours during the following years, dedicating much of his efforts to problems of sericulture, silkworm parasites, and a search for any evidence of the pébrine virus.[36] In his reports, of which copies accompanied a petition to then civil affairs minister Date Munenari, Adams noted the problems of Japan selling its best silkworm cocoons, in addition to specifically stating that there was a need to improve reeling techniques and recommended mechanization of the industry based on a Western model.[37]

In much the same way that Shibusawa and Itō were introduced to Brunat, Hayami sought and received an introduction through the Swiss consul in Yokohama to Casper Mueller, a Swiss silk reeling expert, who had been involved in Italy's silk reeling industry for over 12 years. Following Hayami's guidance and Mueller's advice, with approval from the Ministry of Foreign Affairs (*Gaimushō*), Maebashi domain established the first mechanized reeling facility in Japan based on Western technology in June of 1870.[38] The filature was small, initially with only three reeling frames. It relied on Italian technology and all of the machinery was locally produced. Before the Meiji government had completed construction on Tomioka, the Maebashi filature had moved once and expanded twice. Still based on locally manufactured, wooden, Italian-style machinery, Maebashi filature by June 1871 boasted 36 reeling frames.[39]

The Maebashi facility was neither very profitable nor long lasting, but it was important for providing and disseminating an alternative technology to that employed at Tomioka. The two facilities were built with one goal in common: to serve as models of mechanization and modernization for Japan's local silk reelers. In this sense Maebashi was far more practical than Tomioka; its simpler Italian-style machinery was more within the grasp of Japanese producers – financially, methodologically, and technologically – than Tomioka's French technology. Moreover, it was the 'first attempt to introduce Western reeling techniques as much as possible within an exclusively Japanese context.'[40]

Whereas neither facility's technology was replicated with any significant degree of accuracy, private silk reelers were able to copy the Italian technology far more faithfully than the French. In fact, within five months of its opening, in April 1870, the first group of trainees to visit the model filature arrived at Maebashi. This group of six women were part of the workforce from a newly established filature in Tochigi prefecture that also relied on 'Western-style' machinery that appears to have been based on Italian design. In November of the same year, another group of six women arrived at Hayami's filature for training in Western reeling methods. According to Hayami's recollections, he had additional visitors in June 1871 from Kumamoto prefecture who came to Maebashi to study the machinery and learn mechanized reeling techniques. Similarly, a group from the Shinshū district also came to learn

Maebashi's reeling techniques. Training these women was part of a request that Hayami had received to set up five additional filatures in that area.[41] From the perspective of aiding the basic transfer of technology to local producers, Maebashi would have to be judged more successful than Tomioka.

The first attempt to replicate the government's model filature came in 1874 when a group composed largely of former samurai opened a filature in Nagano prefecture, named Rokkōsha, that was theoretically based solely on Tomioka's French technology. Yet the fact that Rokkōsha was able to serve as a model for others to follow later, betrays the simplicity of the operation. Rokkōsha did not have the government's financial resources, as was the case with Tomioka, and modifications had to be made. These primarily came in the form of cost-saving adaptations to the technology. All of Rokkōsha's reeling frames were locally produced from locally available materials. The copper basins for which Tomioka was famous were replaced with ceramic. Steam engines were replaced by a waterwheel, and kindling fires replaced coal. Yet through all the changes, Rokkōsha still boasted that it produced silk by Tomioka's superior French methods.[42]

Tomioka, Maebashi, and Rokkōsha were visited by silk reelers from all parts of Japan who wished to improve their reeling methods. At the very least, they were curious to see what European technology had to offer and how this could be turned into profit. In many cases, these small-scale producers imported some aspect of what they had seen into their facilities, but even the most painstaking attempts at replicating Tomioka's technologies were far from faithful to the original model. More often than not, firms advertised their silk as being made in a 'Tomioka-style' facility or by 'Tomioka-style' methods after adopting only the most insignificant aspects of the technology.[43] Few, if any, claimed to have modernized through the adoption of Maebashi's technology, although its methods were more widely diffused and its silk had a good reputation and drew high prices in international markets.

Even the small-scale producers who had visited Rokkōsha, but who had never visited Tomioka, and who further simplified the technologies they had seen to match their particular situations, claimed to produce silk by Tomioka's methods. A Saitama prefecture filature established in 1876 that claimed, for example, to be based in part on Tomioka's technology was actually modeled on another small filature located in Seta county.[44] In fact, some merchants stated that reelers *must* go to Tomioka *in person* to inspect the facility for themselves.[45] Arguably done to maximize profit potentials, local reelers who allied themselves with Tomioka had in essence become ambassadors of the government's ideology. By the score, filature owners claimed that their establishments were based on, or modeled after, Tomioka Silk Filature. But in each case, descriptions of the facilities reveal the fallacy of their claims. Few if any used steam reeling techniques, most were small, powered by waterwheels or hand, and many descriptions of 'Tomioka-style filatures' clearly indicate the use of Italian-style reeling frames.[46]

McCallion argues that local reelers were 'simply [being] more politic to make reference to the government's filature.'[47] However, the significance of this attachment is deeper. Tomioka's greatest significance was its mystique. For the government it was a source of domestic and international prestige, for the small producer it was an ideal. The desire to be identified with this ideal was so strong that one Kanagawa filature claimed that their *zaguri* were French-made![48] Tomioka's physical presence represented 'progress and civilization.' While not claiming to be representatives of 'civilization' themselves, small producers became entangled in the belief that association with the Tomioka name alone was enough to imply 'progress' and high quality.[49]

Throughout the first years of the Meiji era, the government was well aware of alternative silk reeling technologies. Its primary focus, however, remained the Tomioka filature and French reeling methods. Attempts within various ministries to develop alternative technologies were largely ignored by the central bureaucracy. In January 1873, the same month that Tomioka began production in earnest, the Ministry of Public Works, Office for the Promotion of Industry (*Kōbushō kankōryō*) broke ground on its own filature in Akasaka, Tokyo.[50] Although the event was marked by a small ceremony, the facility was modest compared to Tomioka. It would receive none of the recognition that accompanied the latter's opening. The entire history of the Akasaka filature is largely a lesson in obscurity. The facility did not even warrant its own paragraph in the ministry's official record, the *Kōbushō enkaku hōkoku*; its treatment is limited to a few sentences mixed in with the ministry's other manufacturing ventures.[51]

Within a month of the ground-breaking ceremony, on 18 February 1873, Akasaka filature's reeling machinery was installed, its workers hired, and working regulations formalized. Since its foreign adviser was none other than Maebashi's Casper Mueller, it is not surprising that the Akasaka filature relied on wooden Italian-style reeling frames and a waterwheel for power.[52] Like Tomioka and Maebashi, the Office for the Promotion of Industry's filature would also serve as a training ground for local silk producers. When Hayami Kenzō visited the facility in March 1873, he noted that the machinery was of good quality and that everything seemed to be functioning in good order.[53]

Unfortunately, the resemblance with the other filatures did not end with its educational function. The Akasaka filature also had financial troubles, which Hayami blamed on managers who were not strict enough with the women reelers.[54] Possibly contributing to the filature's financial woes, and unknown to Hayami, was that operating expenses for the Akasaka filature were lumped together with the Ministry of Public Works' expenditures on the Nagasaki Shipyards and movable-type plant. When this money was exhausted, the Ministry of Finance was often forced to pick up the tab.[55] Given that Tomioka was a joint project by the Ministries of Finance and Civil Affairs, and as we shall later see, a particular minister within the finance ministry disagreed with Hayami regarding the best methods by which to improve the quality of Japanese silk, it is likely that Akasaka's financial difficulties were simply an excuse to abandon the project.

Regardless of true motive, the government wasted no time in divesting itself of the venture. Less than two years after opening its doors, in November 1874, the Akasaka filature was leased to two merchants. It was passed through a series of other private concerns until 1879 when the Ministry of Home Affairs disbanded the entire operation.

Hayami was steadfast in his belief that Japan must manufacture its own reeling machinery as had been done at Akasaka and Maebashi. In a report issued to Commissioner of Revenue (*sozei gonnokami*) Matsukata Masayoshi in October 1873, he argued that domestically produced machines would aid in disseminating new technologies better than imported machinery, and that this would bring greater benefits to the country.[56] This position, however, put Hayami at odds with Shibusawa Eiichi, Matsukata's assistant in the finance ministry's Bureau of Taxation.[57] According to Matsukata's instructions, Hayami and Shibusawa were to formulate a joint report regarding ways in which to improve the manufacture of raw silk. The two could not reach an agreement, however, and in the end Hayami's opinions were presented to Matsukata.[58] In addition to his position on domestically produced reeling machinery, Hayami did not wholly approve of the hiring of expensive foreign advisers. He recommended that the Bureau of Taxation and the Office for the Promotion of Industry hire three 'men of talent' and establish three filatures, each with 100 reeling frames. The 'people,' *kokumin*, would be invited to these filatures where they would become convinced of the merits of Western reeling machinery. Through their efforts at reproducing Western-style machines, start-up costs would be contained. In sum, Hayami urged mutual aid, i.e., filatures of domestic design that would help spread Western technology and methods.[59] For its part, Tomioka would remain unchanged.

Not surprisingly, Hayami was largely correct in his recommendations: the majority of the mechanization of the silk industry, which began in earnest nearly a decade after his report, was based on local initiative and locally produced machines.[60] Despite the claims of many small-time operators, Tomioka's technologies were not that widely disseminated. Throughout its short lifetime, however, the Akasaka filature made significant contributions to the dissemination of practical reeling technologies to local reelers. From the time it opened until the day its machinery was put into storage in a Ministry of Home Affairs warehouse, reelers from all over Japan came to the facility to observe and copy its methods.

A group of nine entrepreneurs from Niigata prefecture visited the Akasaka filature shortly after it opened in 1873. After returning home, they built a facility with 30 frames based on the technology they had studied there. They even improved it by using a steam engine instead of a waterwheel for power.[61] An Ishikawa-based operation also set up a successful filature using 'the silk reeling machines of the Tokyo Kōbushō [Akasaka] filature' in 1875.[62] One Yamanashi prefecture filature, which opened in October 1877, claimed to be based on blending Tomioka's and Akasaka's technologies, although the description is more faithful to an Akasaka copy alone.[63] In Kanagawa prefecture, another filature was established based on Akasaka's machinery, and its reelers were trained at the facility as well.[64] Even as the former Ministry of Public Works filature was being dismantled, its technologies and methods were being adopted. In late 1879, men from Toyoshima county built a filature modeled after the ministry's operation. They even hired the women who had worked there.[65]

International recognition achieved by the Akasaka filature made little difference for its longevity. Only adding an extra dimension of complexity to the issues at hand, this facet of the filature's brief history helps illustrate the government's position, however impractical, of importing and promoting only what it believed were technologies that best exemplified 'progress and civilization.' Tomioka and the Akasaka filature were singled out at the 1873 International Exhibition in Vienna for their high-quality silk. The Austrian official who presented the awards to Japan's representatives stated that 'Tokyo Kankōryō Seishijō [Akasaka] and Tomioka Kankōryō are improved Japanese filatures using Western methods, and are of vital importance [to Japan]. The relative merits of their exports are worthy of praise.' He continued, 'There is a feeling that [these filatures] have advanced along the path of European-style silk reeling methods and to demonstrate that belief, [we are] awarding [these filatures] a Medal of Progress.'[66] Sano Tsunetami, a public works official and proponent of having Japan organize its own national exhibitions, later boasted that Japan and Tomioka did not simply adopt Western methods: 'On the contrary,' he said, 'they *surpassed* them.'[67]

According to Motoyama Yukihiko, the Meiji government wanted to 'supplement its civilization from the outside – what might be called an exterior approach to civilization.'[68] Although the Akasaka filature produced high-quality silk – comparable to what was produced at Tomioka – the physical manifestation of the facility did nothing toward promoting the advancement of 'civilization.' Tomioka, the physical facility more than the silk it produced, was a source of prestige for the government, symbolizing 'progress and civilization' in Japan.

Everything about Tomioka, from its imposing brick facade, to its all-European architecture and centrally located smokestack, to its rows of cast iron machines, steam engines, and gleaming copper basins represented 'progress' and 'civilization' in the Western world order. As Gregory Clancey has argued, stone and brick buildings were a measure of 'civilization': societies without stone ruins or buildings were considered backward by Victorian standards.[69] Similarly, the steam engine, a prominent fixture at Tomioka, was the hallmark of industrial progress in Europe and the United States.[70] The Akasaka filature had none of these attributes.

The government's choice of materials for Tomioka's basic construction and the machinery can be considered impractical and even wasteful from the perspective of technological rationality. Building Tomioka from brick, the blueprint according to one author, copied from the French-designed Yokosuka arsenal, was arguably a good choice as this was a common material from which to build a factory in Europe.[71] However, brick was practically nonexistent in Japan at the time; in the Tomioka area it was a completely unknown commodity. As a result, a good deal of Odaka's and Brunat's efforts went into finding local tile makers and having them formulate a reasonable facsimile, as well as figuring out how to make cement.[72]

The choice of imported cast iron machines was equally problematic. Japan had no iron industry to speak of at the time Tomioka was being constructed. It was, and would remain for decades, difficult if not impossible for local producers to copy the machinery with any degree of accuracy. This condition even forced the government to import its spare parts from abroad. Moreover, the sheer weight of cast iron reeling frames would dictate the use of an inanimate power source. The traditional laborers in the silk industry, women, were not strong enough to turn the frames. As a result, any filature that considered using cast iron machines would also require the use and additional expense of either steam engines or a waterwheel.

Perhaps most telling of the government's attitude is a silk reelers' manual published by the Ministry of Public Works, written by Nagai Yasuoki and Ainé Coye, the former French chief instructor of the ministry's filature the year before the Akasaka filature was permanently closed.[73] Within its three volumes are details on filature construction, where a discussion of the merits of brick buildings figures prominently; and reeling methods, where steam-powered, cast iron, French-style machines are discussed and illustrated.[74] Nagai recommends *only* the use of iron machines stating that they are 'strong and sturdy and will last an eternity.'[75] He claims that with the exception of one or two filatures, undoubtedly Maebashi and Akasaka, silk reeled on wooden machines is of inferior quality.[76] In describing the merits of iron reeling frames, Nagai's terminology is strikingly similar to that used by British architects when describing the merits of stone and brick for buildings and iron for bridges. To many European engineers and architects, 'progress' and 'civilization' were measured by strength and permanence. Wooden machines,

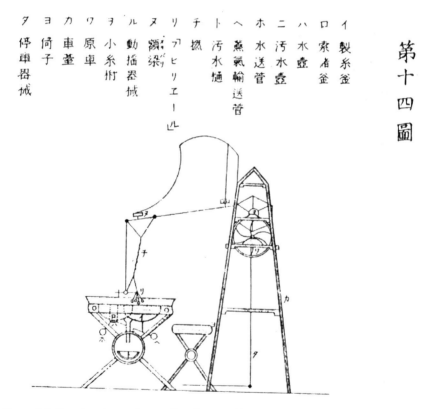

Figure 3.2 French-style cast iron reeling frame and bench from Meiji Era manual (*Source*: Nagai, 1884: n.p.)

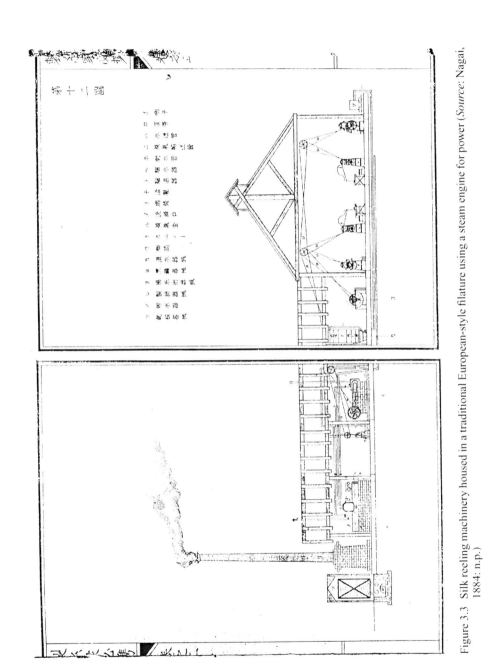

Figure 3.3 Silk reeling machinery housed in a traditional European-style filature using a steam engine for power (*Source*: Nagai, 1884: n.p.)

like wooden buildings, were weak and impermanent, indicative of tradition and a country's backwardness. This view was adopted by Meiji leaders in their quest for proof of Japan's 'progress' in attaining *bunmei kaika* – 'civilization and enlightenment.'[77]

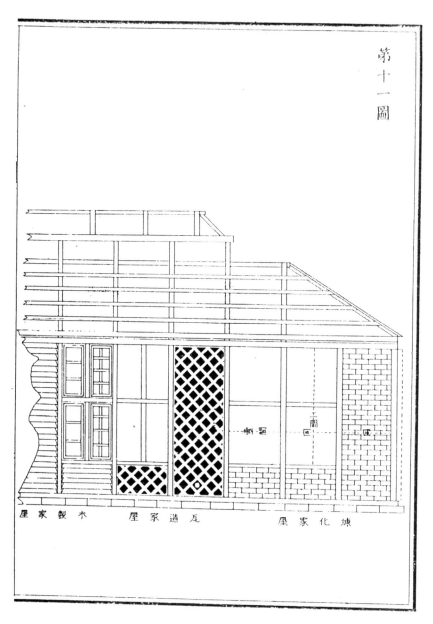

Figure 3.4 Examples of filature exterior materials. Right to left: brick, tile, and wood (*Source*: Nagai, 1884: n.p.)

Nagai's criticisms of Japan's silk reeling industry do not stop with wooden machines: he even finds fault with Japan's European-style machinery. He states that they are too simple compared to the machinery found in Europe, although he never states why additional complexity would be helpful. Interestingly, he also faults Brunat – and even his co-author Coye – for compromising the technologies.[78]

The recommendation of brick building construction ultimately betrays Nagai's beliefs in the material representations of 'progress and civilization.' Nagai again returns to Victorian terminology to describe brick buildings. He states that they are sturdier than wooden or tile veneer buildings, and that they are permanent. He notes that if cost is a problem, one should adopt one of the other building methods, that is wood or tile, but that ultimately a brick building will demonstrate the *'status of the filature.'*[79] Fabricated from wood and housed in a meager wooden structure, yet fully capable of producing silk of equal quality to the silk reeled at Tomioka, Maebashi's and Akasaka's Italian technologies were not promoted in earnest by any ministry within the Meiji government.

Based on a comparison with the Italian technology imported by Hayami at Maebashi, Tomioka was indisputably Japan's premier reeling facility and would retain that title for decades: Tomioka's French technology was decidedly superior. Compared with filatures in France, Tomioka was large and sophisticated, although its orthodox technology had some disadvantages. This problem was understood in Europe. By the mid-nineteenth century many French reelers had recognized the shortcomings of their own techniques and sought to remedy the situation by adopting outside practices. *Croisure*, or crossing, the method by which water is removed from the silk thread to dry and strengthen it, is a perfect example. Traditional French methods relied on a device known as a *chambon*, whereby a number of separate threads were crossed or twisted together. Italian reelers used a mechanical device called a *tavelle* that used a number of small pulleys or wheels to cross a single strand of thread over itself. The former method has the disadvantages of being technically and operationally complex. It also limits the number of threads reeled per reeling frame to four, which translates into the maximum number of threads per reeler. In short, this method, in addition to being complicated, limits output and therefore the potential for future expanded production. The Italian *tavelle* method has none of these disadvantages and was adopted by many French filatures.[80]

In the Ministry of Public Works' reeling handbooks, both methods are described, but only the French method is illustrated. While stating that the method in the picture is good, the author points out its problems, that is, the difficulty of its mastery. When describing and detailing the merits of the methodologically simpler *tavelle*, however, Nagai offers no judgement.[81]

Choice of technique: an alternative theory

If Tomioka was neither technologically rational nor economically sound, it had to have been established on some other basis. McCallion has correctly argued that part of Tomioka's problem was that the government never had a clear vision, let alone plans, for the facility. Tomioka's proposed mission was for it to serve as a

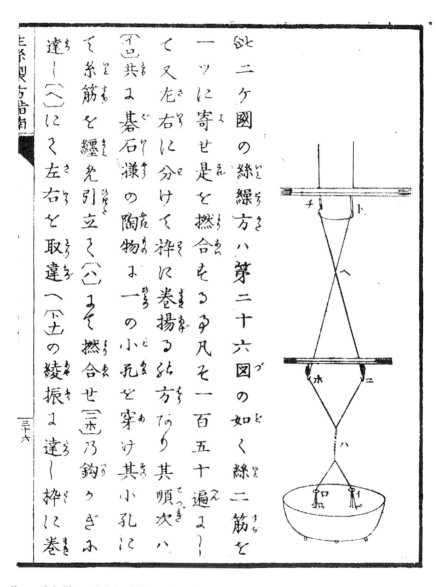

Figure 3.5 Figure 3.5 (and Figure 3.6) shows the *chambon* method of *croisure*. Starting at the bottom, two sets of silk filaments are twisted together into two threads. They are intertwined and then separated. This action strips excess water, cleans, and binds the individual filaments into a single thread. The two threads then pass over separate loops from where each is reeled onto a separate take-up reel. This method, adopted at Tomioka, is also know as *toyomori* (*Source*: Tachi, 1874: 36)

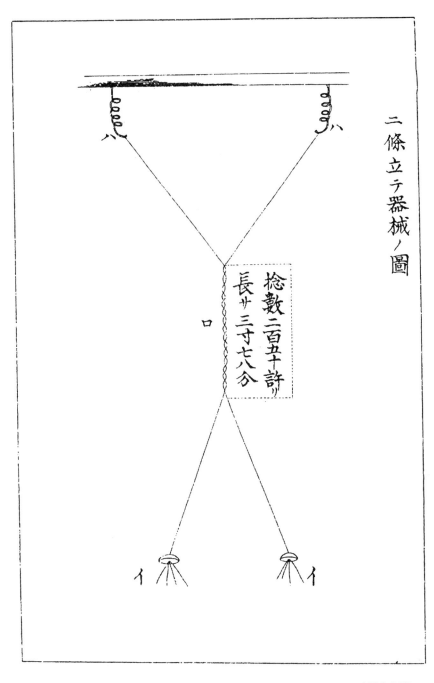

Figure 3.6 *Chambon* (*toyomori*) method of *croisure* (*Source*: Itō Moemon, 1886: 549)

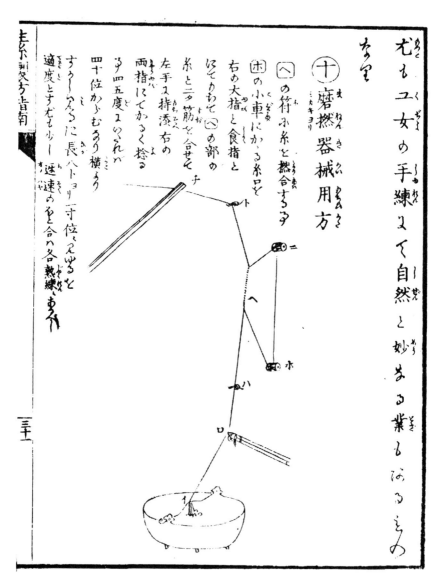

Figure 3.7 Figure 3.7 (and Figure 3.8) are of a *tavelle*. Several silk filaments are gathered together, fed through a loop of hair, wound through a number of small guide wheels and twisted over itself in order to strip water from and strengthen the silk thread. This method, also known as *kenneru*, was used at Maebashi and Asasaka filatures (*Source*: Tachi, 1874: 31)

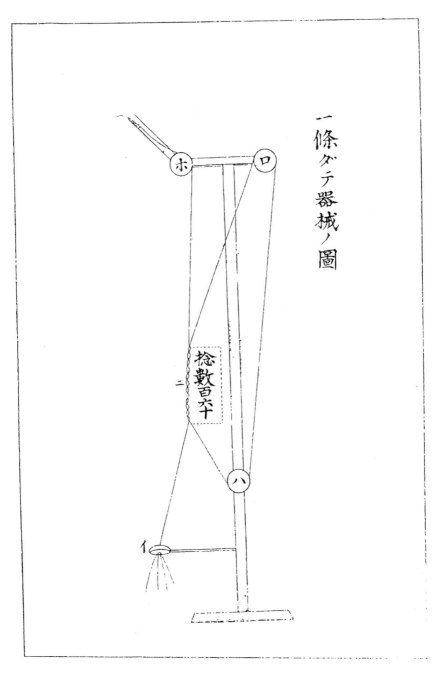

Figure 3.8 *Tavelle* (*Kenneru*) method of *croisure* (source: Itō Moemon, 1886: 547)

model of industrialization that private reeling concerns could follow to improve the quality of Japan's raw silk. At other times, Tomioka also was supposed to operate as a profitable business venture. Whereas these goals should not be mutually exclusive, they were problematic for the Meiji government. From what evidence remains, however, this need not necessarily have been the case.

It appears that the Ministries of Finance's and Civil Affairs' original ideas for a publicly operated, model reeling facility were more in accord with what Brunat had suggested, but that somewhere along the line, ideas about Tomioka's purpose changed. Odaka Atsutada stated that he, Sugiura, and Brunat traveled to Maebashi some time around July 1870 to visit Hayami Kenzō and ask about Mueller's plans for the facility. He noted that at the time, Mueller's plans for Maebashi were more or less the same as what the government had in mind for Tomioka. At least according to these recollections, the Meiji government originally did not anticipate their first entrepreneurial venture into the silk industry to be anything more than a relatively humble facility.[82]

Shibusawa Eiichi is traditionally credited with establishing the Tomioka filature. And although the official record presents no direct evidence as to whose suggestions led the Japanese government literally to import a French filature, there are strong indications that both Shibusawa Eiichi and Sugiura Yuzuru were behind the decision. In an address given in the late-1880s evaluating the progress of Japan's silk reeling industry, Odaka Atsutada stated that the Ministry of Finance used Shibusawa's proposal for Tomioka.[83] Odaka's praise of Shibusawa and Shibusawa's comments regarding traditional Japanese silk reeling methods are also revealing. Shibusawa was a strong critic of Japanese silk reeling techniques. He characterized them as imperfect and recommended the adoption of European methods. Odaka quoted Shibusawa's earlier statement that 'the Japanese way of reeling silk was deficient (compared to that of the West).'[84] Praising the progress made by Japan's silk reeling industry in recent decades, Odaka noted that the foundations of Japan's modern industry relied on the efforts of Shibusawa Eiichi.[85] The question then becomes what inspired Shibusawa to change the plans for a relatively humble model factory into an ideal to which few could aspire and none could imitate.

In 1867, Shibusawa Eiichi and Sugiura Yuzuru were part of a Tokugawa mission to Europe where they had the opportunity to visit the Fifth International Exhibition in Paris.[86] Ultimately Shibusawa would live in Paris for six months serving as an attendant for Tokugawa Akitake, the shogun's younger brother.[87] This was the same Tokugawa mission that Alex von Siebold accompanied to Paris, providing regular commentary to the British legation in Yokohama.

Traveling to Lyons by train, Shibusawa and Sugiura had only a brief opportunity for sightseeing, but were impressed with the city nonetheless. Taking in the city's flavor, its gardens, stores, and homes, the pair noted the spaciousness and cosmopolitan character of Lyons. The two were particularly keen on the abundant variety and quality of women's silk accessories. And although Shibusawa and Sugiura saw numerous merchants selling the machinery and appliances of the silk trades on their tour of the city, neither man had the opportunity to visit a reeling or weaving facility.[88] They were nonetheless able to examine Lyons silk more closely at the Paris exhibition in June. There, both Shibusawa and Sugiura were impressed, not only by the quality of Lyons-made silk fabrics, but by the sheer magnitude of the exhibition – not to mention Paris itself. Both men noted the prestige attached to winning awards at such an exhibition and how countries vied for distinction by displaying their most advanced and finest products. In addition to silk, they were especially impressed with the cast iron machinery and steam engines.[89]

This was not Sugiura's first trip abroad. From a young age, based on his studies, he had developed a positive attitude toward Western science.[90] He later had the opportunity to visit Europe, first in 1861 and again in 1863, as a Foreign Magistrate (*gaikoku bugyō*).[91] Steam-powered printing with lead type seems to have made the biggest impression on Sugiura as he equated communications fostered by steam power as the root of human wisdom and 'civilization.'[92] Based on his observations in France in 1863, Sugiura became a proponent of importing Western 'civilization' to Japan.[93] His return to Paris in 1866–67 with Shibusawa and his visit to the International Exhibition only served to develop further his ideas and dedication to building a Western-style public society in Japan.[94] More than his previous visits to newspapers, hospitals, and factories, it was the Paris exhibition that widened Sugiura's perception and understanding of 'progress and civilization' in the West. And it was through their time together in Paris that Sugiura and Shibusawa became trusting friends who shared similar visions for the future of Japan.[95]

After the exhibition ended, the two toured Switzerland, Italy, Great Britain, and the Netherlands. In September, Shibusawa and Sugiura visited filatures and weaving establishments in Switzerland which they described as being merely artisanal or handicraft in nature. While commenting favorably on the detail of some of the fabrics, the two were obviously less impressed with the quality of what they had seen and with the small size of the facilities they had visited. Their travels through Italy, while including a trek through the mountains in a horse-drawn carriage, a far cry from the steam locomotives to which they had become accustomed, did not include a visit to a filature. Silk, while prominent in the narrative of their travels in France, was not even mentioned with regard to Italy.[96] This is most surprising, given that at the time Italy was the world's largest producer of fine silk fabrics and a player in the Yokohama silk trade. A few things about Italy made an impression on Shibusawa, however. He described the carriage in which they traveled in detail, noting that it was the means of transportation in Europe used *before* the invention of the train: one of the nineteenth century's most identifiable measures of a country's level of 'civilization.'[97]

It is likely that this impression of Western weaving and reeling methods, coupled with their awe of 'civilization' as seen in Paris and Lyons, further biased Shibusawa's and Sugiura's opinions in favor of French reeling techniques. For years, French officials had been trying to cultivate France's image in Japan as Europe's center of 'civilization and enlightenment,' because the primary concern of the French government regarding Japan was influence and prestige.[98] At the time, Shibusawa and Sugiura were in Europe under the authority of the Tokugawa *bakufu*. It is possible that the relationship that Leon Roches had carefully cultivated during the mid-1860s saw fruition in the importation of French silk reeling technology under the guidance of these former *bakufu* men who served the new Meiji government.

While Japanese silk dealers and government officials recognized the quality of French silk, they were apparently uninformed about the process by which it was made.[99] Throughout France, local silk reelers had long been modifying their technologies with Italian designs. By the mid-nineteenth century, Italian reeling methods were widespread throughout France and many silk reelers considered them superior to the traditional French methods.[100] In all likelihood, any silk reeling machinery that Shibusawa and Sugiura had seen in Lyons was probably of Italian design, given that that city was one of the first to utilize the superior Italian methods.[101] It would only be in the years which followed that Meiji officials would come to recognize, or perhaps accept, the position of Italian silk. Even if the government's representatives did not understand the technical details of French and Italian silk reeling machinery, they certainly recognized the political and economic importance of the two countries for Japan.

Diplomatic relations with Italy were formally established in 1866; commercial relations in the form of the raw silk and silkworm egg trades through Yokohama were established shortly thereafter.[102] Much for the same reasons that the French were forced to look outside their borders for raw silk and silkworms, the pébrine virus also brought Italian silk merchants to Japan. The disease had much more serious consequences for France's industry, however, and the volume of trade with Italy was correspondingly lower. As noted earlier, not only did Shibusawa's trip to Italy bypass silk reeling facilities, years later the Iwakura Mission also relegated the industry in general to an unofficial side-trip. Indeed, our knowledge of the Mission's silk-related activities comes from the Italian press, not the official record of the Japanese visitors to Italy.[103] Although Japanese sericulturists were later sent to Italy to study scientific methods of silkworm rearing, Italy was seen more as a source of competition, as was indicated in an 1873 report by the Japanese consul-general to Italy who noted that the Italian government was licensing trading companies for the purpose of promoting silkworm rearing and eliminating the imports of Japanese silkworm egg cards.[104] Shibusawa also seems to have considered Italy a source of competition. In the summer of 1873 he was again in Italy, this time to gauge the quality of Tomioka's silk against the Italian product. He happily reported that the two were of 'equal quality' and that the people of Italy appeared to be impressed. Tomioka's silk had brought Japan prestige and glory, he reported, and all that remained was to 'await our products being labeled the finest in the world.'[105]

Possibly more crucial for the new Meiji government and its selection of reeling technology, however, was their evaluation of the two countries' political importance. Shibusawa noted in his travel log for 23 October 1867 that Italy had been torn by civil war a year before the Tokugawa mission arrived. He also noted that Tokugawa Akitake's audience with King Victor Emmanuel II was cut short because of disturbances in Rome. In fact, Rome would not officially be part of the unified Kingdom of Italy until 1870. While Shibusawa praised the skill of the artisans at a mosaic factory, these comments pale in comparison to the traveler's evaluations of French manufacturing and Paris.[106] Much like the low level of American diplomatic participation in Japan following the U. S. Civil War, Italy was similarly relatively inactive in early Meiji politics – especially when compared to Britain and France. Although Italy made contributions to Japan's military technology during the Meiji period, these efforts were insignificant in comparison to Roches' exploits and French involvement in general. Italy's greatest contributions were in the introduction of Western art and music in the late 1870s.

As a result of France's greater participation in Japanese politics and trade, the impressions of the country's technical and political prestige formed at the Fifth International Exhibition in Paris, and the aura of 'civilization,' which for Shibusawa and Sugiura seemed to exude from every feature of French society, the Meiji government decided to import strictly orthodox French silk reeling technology for its model filature at Tomioka. There was no technical evaluation of the machinery, nor was there any attempt to ascertain whether this new technology would be appropriate for Japan or Japanese silk reelers.[107]

Validation and idealism

Determining choice of technique is a difficult process regardless of time or location. In Meiji Japan, the task was further complicated by the nature of the government, internal disturbances, and external conditions, such as the unequal treaties and the need to demonstrate the legitimacy of the new Imperial government to the leading Western powers. That the government knew of alternative Western reeling technologies seemingly adds further complexity of the debate surrounding the decision to import orthodox French technology. A third unspoken function for Tomioka Filature explains why the government deviated from its original plans and, more importantly, explains why government officials ignored expert advice and attempted to transfer a technology without considering factors such as those discussed above. Tomioka was an ideal. It was designed to be the physical manifestation of the (new) central government's authority over the domain, and it was to serve as an exemplar of Western 'civilization' in Japan. In the quest for 'progress and civilization,' whose model better to follow than France (and Britain), the key political and economic players, who were also the archetypes of *bunmei kaika* materiality?

At the time Tomioka was conceived, the Meiji government's political position was anything but assured. When Geisenheimer first approached Itō in 1869 with the suggestion of establishing a filature using Western technology, the *daimyō* still had not, even symbolically, 'returned' their domains to the Emperor. In 1870, when Hayami established the Maebashi filature, it was under the authority of Maebashi domain, not the central government. In fact, the domains were abolished six months *after* construction on Tomioka began. Umegaki Michio argues that the first few years following the restoration of imperial rule were marked by a series of political moves that could best be described as a 'simultaneous dispersion and consolidation of political power.'[108] In an effort to stabilize the country, members of the new government on the one hand sought to consolidate its political power vis-à-vis the *bakuhan* (*bakufu* and domain) system, and on the other hand sought to invest the domains with constitutional equality in order to eliminate inter-domain rivalries.

The new government's change in plans for Tomioka reflected its need politically to situate itself above the domains, and more importantly, the now defunct *bakufu*. After all, Western-style factories had been part of Satsuma domain's industrial landscape as part of its efforts to industrialize the cotton industry in the decade before the Restoration; and it was under Tokugawa authority in 1865 that French engineers built the Yokosuka ironworks, Japan's first official attempt to import Western industrial technology under the supervision of foreign advisers. Toward that end, any industrial venture sponsored by the new government would also serve the purpose of legitimizing its political authority.

The Meiji government also needed to demonstrate that it was the legitimate heir to Tokugawa political authority to the Western powers whose highly visible commercial and quasi diplomatic activities in the treaty ports kept the bureaucracy on edge. From the time the unequal treaties were signed, first the *bakufu*, various domain, and then the new imperial government recognized the necessity of demonstrating their political authority to the Western powers. For their part, French and British officials in Yokohama continually speculated as to who truly held power in Japan. And although the struggles of the dying *bakufu* were typically interpreted as attempts to 'invalidate the legal basis of the treaties,'[109] the problem was not simply a matter of assuring the West that the Imperial government was in control. It was also trying to demonstrate to the West that Japan was indeed 'civilized.' On the popular level, 'civilization' appeared in Japan through the adoption of Western clothes, hairstyles, and other insignificant, although easily observed and criticized, gestures.[110]

For the government, however well dressed its members may have been, the task was greater. By importing a full-blown modern French factory, the new government believed it was importing Western civilization to Japan. The Ministry of Civil Affairs was also not alone in its belief that adopting the material culture of the West would be representative of 'civilization and progress' in the country. Writing in the summer of 1872, Inoue Kaoru, then head of the finance ministry, expressed his frustration to Kido Takayoshi about the economics of this type of thinking. He complained about excessive capital expenditures: 'regardless of the manner by which we choose to raise [our] level of civilization, there will not be any way to make use of it.'[111] The following January he again made clear his frustration to Kido: 'Although the Finance Department alone has tried to limit government spending, the other departments, with the idea of gaining parity with the West, are insisting on [their] positive policies.'[112] Hayami Kenzō expressed his doubts as to the practical benefits of the government's insistence on importing material civilization: 'The results for those people who intend to bring civilization to the hearts of the Japanese can either cause great harm or be very beneficial.'[113] Hayami urged that the government not import its machinery and use Japanese-made alternatives instead. He argued that following this plan would make possible both high-quality silk and the dissemination of better reeling technologies.[114]

At some point between July and November 1870, something changed in the government's thinking about what type of facility Tomioka was to be. The alternative technologies, like those found at Maebashi and Akasaka, were capable of producing high-quality silk. However, their use of traditional materials, that is wood, and unsophisticated appearance made them unacceptable from an ideological perspective. During his visit to Maebashi filature in July 1870, Odaka betrayed this very point. While impressed with the quality of the silk, he expressed dismay at the 'rickety' wooden machines of which Mueller and Hayami appeared to be so proud.[115] If one's mission was to raise the level of 'civilization' in Japan vis-à-vis the West, then only the most 'modern' facilities made from the most 'modern' materials were acceptable.

Economic and technological considerations do not seem to have been a concern. After all, Odaka and Brunat had no doubts about the abilities of Maebashi's alternative technologies to produce high-quality silk. Nor were they concerned about the ability of home-made reeling apparatus to be successful. As part of their evaluation of Maebashi's technology, Odaka and Brunat had four experienced workers reel silk for 30 days on machines described by Odaka as 'Japanese-style modified to the least extent possible to make them European-style.' At the end of the trial period the machines and silk were said to be of high quality and equal to those of Maebashi.[116] For Odaka and Brunat, economics were a consideration. They were both concerned about the expenses involved with establishing the model filature, although their caution and worries were disregarded higher up the line.

Shibusawa's retrospective appraisals of Tomioka are indicative of beliefs in the superiority of Tomioka's technology and the facility's greatest value. They illustrate what one may argue was *the* determining factor in Tomioka's design: ideology. Writing in the mid-1930s, Shibusawa stated:

> Although there may be some parallel example [of a reeling facility], nothing rivaled Tomioka in terms of being on such a grand scale and so perfect and complete. Its reputation rose to the world's attention and it was admired as a model facility from thriving cities to the countryside.[117]

Shibusawa's comments are similar to those of government officials who were opposed to selling the facility during the Matsukata Deflation. In 1881, a bureaucrat in the Ministry of Agriculture and Commerce argued that Tomioka must be maintained because it was a source of international prestige for the Meiji government. As early as 1873 this was also the case. Sano Tsunetami's report on that year's International Exhibition in Vienna clearly describes the honor and prestige which the medals won by the model filatures had brought to Japan. He stated that the 'kankōryō [Tomioka and Akasaka] filatures' silk is of the best quality and to look at its thread is to *see progress*.'[118] Time and again, government officials ignored, or at least rationalized, Tomioka's financial and operational difficulties because of its prestige value. With this prestige came the belief that Japan's industrial progress as embodied in Tomioka was helping the country become 'civilized.'

Conclusion

When Tomioka Silk Filature opened its doors in 1872, it was the product of neither careful technical nor economic evaluation. More accurately, alternative technologies

Table 3.2 Commendations received by Japan at 1873 International Exhibition, Vienna, Austria

Medals of honor	Silk	Textiles	Total
Progress Medals	2	2	4
Meritorious Service	2	5	7
Joint Praise	0	4	4
Special Commendation	3	11	14

Source: Sano Tsunetami, 1875: vol. 1, section 6: 8

were evaluated and the costs were considered, however, the advice the government received was largely ignored. It appears that at its inception Tomioka was to be a modest venture, probably nothing more elaborate than the small mill Hayami Kenzō had built at Maebashi. This is also largely reflected in Brunat's original proposal to the government. Brunat and Odaka evaluated Maebashi's technologies and were favorably impressed with the quality of the silk it produced. They even conducted their own experiments on locally produced machines to see if they could duplicate or surpass what was being done at the former domain-based enterprise.

Brunat's proposal, while arguably written from the dual perspectives of establishing a profitable business and improving the quality of Japan's raw silk, considered any number of factors that reflected his training in Lyons and the state of the industry in general. He recommended modifying local technologies with recent European innovations only to the extent that it would not disrupt the abilities of local reelers to function in their craft. At the same time, if his recommendations were followed, it would have brought Japan's silk industry to the level of many European reeling facilities.[119] This was also in accord with a directive issued by the Ministry of Civil Affairs in February 1870, which called for reelers and merchants to set aside their greed and build European-style reeling machines in an effort to improve the quality of Japan's raw silk.[120]

In the months which followed the decision to build the facility, the government's thinking changed. Not only would Tomioka serve as a model of mechanization for local reelers, but it would, more importantly, be an exemplar of 'civilization' in Japan. Along with caution, Brunat's prospectus was abandoned as he was ordered to import and build a full-blown French filature in the backwoods of Gunma prefecture. From what evidence remains, this decision was based on the recommendations of Shibusawa Eiichi, who had been favorably impressed with France, Lyons silk, and the prestige it brought that nation at the Fifth International Exhibition in Paris in the spring of 1867.

The government's lack of caution, its alternative basis for choice of technique, was largely a function of the times. In an effort to assert its position over the domain, the new government needed a model facility that placed it well above these semi-autonomous territories which were still vying for political influence. The new Imperial government also needed to demonstrate to the international community that it was the legitimate heir to Tokugawa political authority. Toward this end, it used the material trappings of the West to its best advantage. Tomioka Silk Filature would represent the march of Western 'progress' and 'civilization' in Japan. For the same reasons that the Ginza was completely rebuilt in brick after an 1872 fire, as evidence of 'civilization' and, to quote Saigō Takamori, 'for the sake of our honor,' Tomioka was designed on the basis of symbolism.[121]

The Meiji government's first industrial venture was not part of any grand plan. It was based on a process in which the components of mechanization were gradually incorporated and assimilated to accommodate the government's varied demands. The primacy of technological development and improving the quality of raw silk gave way to economics; economics succumbed to ideology. In the end, the new government had its filature, however ill-defined and unstated its purpose. Because Tomioka and the government's efforts are frequently evaluated in terms of financial success, the facility is often judged a failure. However, this myopic view of Japanese industrialization fails to take ideology into account, and was, I believe, the fundamental factor in the choice of technique. As the Tomioka model demonstrates, beliefs in the ability of transferred technological artifacts to impart cultural and social values are critical to their selection.

4 Smelting for civilization

Technological choice and the modernization of the iron industry

In early 1872, with hopes of rousing the Meiji government to action, Vice Minister of Public Works (*Kōbushō dayū*) Yamao Yōzō submitted a petition to the Central Chamber (*Sei'in*) regarding the promotion of Japan's mining industries. The gist of his proposal was that mining was the basis of a '*rich nation.*' Through the promotion of mining, the government could provide machinery to the people, build factories, and awaken the nation's 'spirit of large industry.' Just in case the Chamber needed further persuasion, Yamao offered concrete economic incentives. He pointed to reports in the British press that the price of copper – an abundant commodity in Japan – had increased by more than 25 percent in the past few months.[1] Within a few years of Yamao's petition, the government was fully involved with the promotion of mining and metallurgy, opening mines and smelting facilities with the hope of exploiting Japan's copper, silver, gold, and, of course, iron reserves.

A quick look at this list of metals, analogous to media of exchange, is revealing of early government attitudes toward iron.[2] Copper, silver, and gold, along with coal – black gold? – were valuable commodities readily convertible on the international market. When not bartering with silk or tea, the new Meiji government needed these metals to support its various attempts at 'civilization' building. Foreign speculators also demanded Japanese copper, silver, and gold, while the great merchant houses, shipping lines, and navies of Europe and the United States purchased Japanese coal for their steamers.

By and large, iron was outside the financial equation. If, however, one wanted 'sturdy' machines and 'permanent' bridges, not to mention a railroad and modern navy, iron was essential. But iron was a material of paradox. While not being a material *of* 'civilization,' it was the material with which one *built* 'civilization.' When construction was completed on an iron bridge over the Nakajima River in Nagasaki, for example, everyone was amazed at 'Inoue's iron bridge.' Despite the fact that Japan had neither sufficient resources nor capital, the man responsible for having it erected, Inoue Kaoru, a proponent of bringing 'civilization' to Japan, insisted on building with iron. Inoue was keen on iron's cultural implications, adopting the language of British architects and officials when he observed that the iron bridge was far 'stronger and more permanent' that any traditional wooden structure.[3]

A similar project was undertaken two years later in Yokohama. Japan's second iron bridge, which cost the government over US$15,000, (approximately ¥30,000) was also the subject of much controversy. Critics argued that Japan could ill-afford the luxury of this type of structure. Proponents, including British engineers and government officials, however, called it representative of the 'spirit of Japanese progress.' Iron bridges would represent 'progress' and 'civilization' in Japan.[4]

Sano Tsunetami similarly described how building an iron industry would help bring 'civilization' to Japan. Looking at the material culture of 'civilizations – tools, weapons, and iron clad warships' – from the vantage point of the 1873 International Exhibition in Vienna, he identified iron as the material which would help Japan attain recognition in the West.[5] Identifying the reported mineral wealth of the Rikuchū Heigori iron mines, Sano stated that the amount the country spent on building an ironworks would bear directly on helping Japan 'attain its chosen level of "civilization."'[6] He did not ignore Japan's other minerals, citing copper as second in importance when building 'civilization.' Daily use items are fashioned from copper, he stated, and Japan should endeavor to increase exports of copper and brass, as well as coal. Iron was a product for import substitution, and one from which the nation could build its 'modern' textile machinery.[7]

Like the government's promotion of the silk reeling industry, the development of the iron industry was affected by numerous factors that have rarely been considered crucial to the process. In fact, most studies that discuss the early years of the Meiji iron industry are rather narrow in perspective, only considering import substitution or unsupportable hypotheses of slavish devotion to the whims of foreign advisers as critical.[8] Interestingly, my case studies of the iron and silk industries may be considered mirror-images. Whereas foreign involvement was nearly constant and significantly contributed to the government's course of action on behalf of silk, foreign governments were largely inactive in Japan's iron industry, and the government's need to prove Japan's mettle in this area was correspondingly low. As demonstrated in the previous chapter, choosing the most modern technologies for Tomioka filature was technologically and economically inappropriate. Selection of the most modern methods for Kamaishi Ironworks, while probably appropriate from a technological perspective, was supported by faulty reasoning. The nature of the two government ventures was also very different. Tomioka was supposed to be a model factory, and it was under constant public scrutiny. The goings-on at Kamaishi Ironworks, although a public venture, were rarely publicized. It was not a model factory where metallurgists-to-be and entrepreneurs could come to witness the latest Western smelting technologies. Moreover, and perhaps most importantly, Kamaishi carried none of the prestige that was attached to the model filature, and as such it was relatively easy for the government to abandon the project when things did not go according to plan.

Kamaishi Ironworks: the early years

The story behind the government's plan to build an ironworks near Kamaishi village in Iwate prefecture, the problems associated with the project, and its sale in 1883 during the Matsukata Deflation are assumed to be well known.[9] Familiarity in this case, however, has bred inaccuracies that have become magnified with time. For this reason, I will start from the beginning in the hope of providing a more accurate account of the development of the government's ironworks.[10]

Kamaishi's Bakumatsu–Meiji era history can be roughly divided into three periods: from the time Ōshima Takatō constructed the first Western-style blast furnace at Kamaishi until the establishment of the Ministry of Public Works-run facility (1854–74); the public works era (1874–83); and its time as Kamaishi Mine, Tanaka Ironworks (1883–1924).[11] In describing the period prior to the Meiji government's involvement at Kamaishi, I will briefly outline the foundation of Japan's iron industry based on Western technology in an attempt to set the scene for later events which shaped the course of Kamaishi as a state-run enterprise.

Ōshima Takatō, considered the father of Japan's modern iron industry, is the person credited with building and operating the first Western-style blast furnace using iron ore in Japan. He was the son of a mid-level samurai from Morioka, Nambu-domain (present day Iwate prefecture) who, at the age of 17 was ordered to Edo (Tokyo) where he studied Dutch Learning (*rangaku*).[12] Three years later, in 1847, he was sent to Nagasaki to learn Dutch medicine. More interested in military technology and cannon founding than medicine, however, Ōshima spent his days in Nagasaki studying military science, metallurgy, and mining.[13] Among Ōshima's achievements at this early stage of his career was the publication of *Seiyō tekkō chuzō hen* (*A Treatise on Casting Western Iron Cannon*), a joint translation of Ulrich Huguenin's 1826 text *Het Gietwezen in 'sRijks Ijzer-Geschutgieterij, te Luik* (*Iron Casting at the Rijks State-owned Cannon Works at Liuk*) in which he described Western methods for producing cast iron artillery.[14]

After returning to Morioka in the early 1850s, Ōshima approached Nambu domain officials for permission to construct a Western-style furnace in the Kamaishi area. His intention was to produce iron for casting cannon. After some initial prospecting, and with financial backing from Ohara Zengorō, a Morioka merchant, Ōshima produced a few pigs of iron by 1854. The project was under-capitalized, however, and the operation was abandoned after Ohara ran out of money.[15] As fate would have it, Ōshima had a second chance. He was summoned to Mito-domain (present day Ibaragi prefecture) where he was invited to supervise the development of reverberatory furnaces for that domain. Mito, a collateral Tokugawa house, would become a hotbed of anti-foreign activity, and Ōshima's efforts at producing cast iron cannon were in concert with that domain's anti-foreign (*jōi*) policies.[16] In late November 1855, the Mito furnace, located in Nakaminato, Mito's capital (*jōkamachi*), was put into operation, and by March of the following year, the operation was prematurely declared a success.[17] Within months, however, Ōshima and his associates began noticing significant problems with the pig iron they were using.[18] Because the pig iron used in the Mito furnace was made from iron sand, the traditional raw material of Japanese iron production, its irregular composition resulted in uneven reduction and a final product of mixed quality. Although successful at casting mortar from the iron produced at the Mito reverberatory furnaces, it was necessary to remedy the quality problem.[19]

Having been granted 100 days' leave of absence by Mito officials, Ōshima returned to Nambu to once again seek permission to mine the region's iron ore and build an ironworks at Ōhashi in the Rikuchū region. The chief retainer of Nambu granted Ōshima's request in November 1856. Within four months he began furnace construction, and before the end of the year, on 1 December 1857, iron began to flow from Ōshima's blast furnace.[20]

This success did not come without a price, however. Shortly after Ōshima had arrived at Ōhashi in 1856, a severe storm stuck Nakaminato.[21] Strong winds damaged one of Ōshima's Mito reverberatory furnaces – toppling the chimney – and casting operations were suspended for a number of months while the furnace was repaired. Although casting cannon continued at the domain's second furnace, Ōshima appears to have been under constant pressure to make sure that the facility was up and running at full capacity as quickly as possible, all the while remaining at Ōhashi engaged in blast furnace construction.[22] The Mito furnace was back on line by February 1857, but this event seems to have left its mark on Ōshima. As will be demonstrated, Ōshima's primary concern – to the detriment of all others – when proposing designs for the Meiji government's ironworks at Kamaishi, was finding a location that was sheltered from the elements.

Smelting at Ōhashi continued on a routine basis. The iron produced by this private enterprise was regularly sent to Mito for that domain's cannon casting project. At Ōshima's suggestion, in May 1858, the entire Ōhashi operation (mine and furnaces) was turned over to Nambu domain. As a public venture the facility was equally successful; its iron even being shipped to Edo as payment for repairs on Edo castle.[23] The Mito furnaces were closed as a result of Tokugawa Nariakira's unrelenting anti-foreign position and subsequent confinement by the *bakufu*. Released from his obligations to that domain, Ōshima continued to serve as an inspector at Ōhashi. He was appointed head technologist at Kamaishi in 1859, and remained at this post throughout the 1860s, expanding operations and erecting additional furnaces at Ōhashi, Hashino, Sahinai, Kuribayashi, and Sunagowatari. By the time the new Meiji government decided to build its ironworks at Kamaishi in 1873, there were a dozen furnaces in all.[24]

Table 4.1 Kamaishi area furnaces, c. 1874

Location	Number	Year built	Approximate annual production
Ōhashi	5	1857–3 1872–2	1,000 tons
Hashino	3	1858	860 tons
Sahinai	2	1859	344 tons
Kuribayashi	2	1868	344 tons
Sunagowatari	1	1865	2 tons[b]
Katsushi	1	1865–8[a]	2 tons[b]

Source: Based on Mori, 1957: 77; Fuji Seitetsu Kabushikigaisha, 9, 'Watanabe,' 1992: 109–120

Notes:
[a] Date given as *Keio nenkan*
[b] daily calculations based on Godfrey's observations

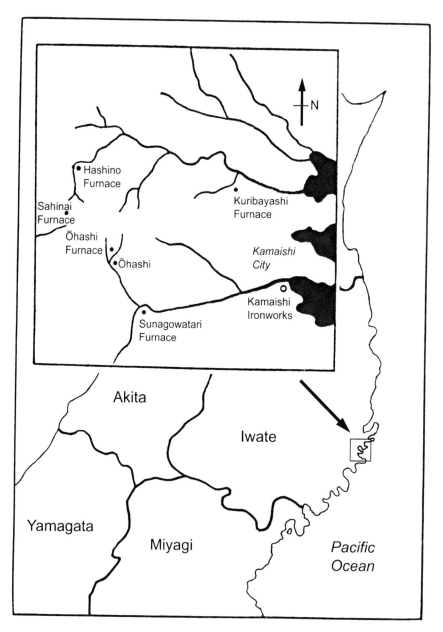

Map 4.1 Location of blast furnaces in Kamaishi area (Based on Mori and Itabashi, 1957:
75)

Kamaishi Ironworks: The Kōbushō years

If any one person was familiar with the Kamaishi area and its productive iron capacities, it was Ōshima Takatō. However, the decision to build an ironworks at Kamaishi was based on recommendations in a foreign adviser's report. Hired in September 1871, J. G. H. Godfrey, Chief Mining Engineer for the Meiji government, toured the Tōhoku region of northern Japan the following summer.[25] Stopping to inspect the iron mines in the Heigori district of Rikuchū, the region where Kamaishi is located, he noted that the supply and quality of the iron ore was good and that working these mines should be profitable.[26] Ōshima was not present at Kamaishi when Godfrey visited, nor was he even part of the original decision-making process. As a member of the Iwakura Mission, Ōshima was in London at the time of Godfrey's tour.[27] He returned to Japan in July 1873 and quickly toured the Kamaishi area with the head of the public works ministry's mining section, Yoshii Tōru, just prior to the government's official announcement to appropriate the mines later that same month.[28]

It may seem ironic that Ōshima, the man responsible for first exploiting and developing the Kamaishi region's iron producing potential, was not included in the selection process. The government had great aspirations for this project and perhaps it was the fact that their facility was to be constructed on such a different scale from anything Ōshima could have even imagined that led to him being somewhat marginalized throughout the process.

During his tour, Godfrey scrutinized every detail of iron production in the region.[29] He was a keen observer who commented on everything from the price of firebricks, to the number of laborers at each operation and how much they were paid. He noted the amount of fuel, ore, and limestone flux used at each facility, the distances between furnace, mine, and market, and detailed costs along the way. When Godfrey arrived in the Kamaishi area, only the blast furnaces at Ōhashi and Hashino were still in operation, and then only on occasion. He observed that there were a number of problems with production including the expense and difficulty of transporting raw materials through the mountainous terrain – compounded by a lack of roads. A scarcity of charcoal in the area surrounding the Hashino furnace site, for example, combined with the difficulty and expense of transportation, meant that only two of the three blast furnaces could be kept in operation at any one time. Hashino's ironworkers similarly abandoned the practice of using limestone flux primarily because of the distance from which it had to be imported.[30]

There were also technical problems with these early furnaces. Godfrey noted the importance of finding a better quality fire clay from which to construct the furnace lining. Without a better lining material, he stated, in general, the furnaces could only be operated for short periods of time before repair was necessary. Taking Hashino as representative, Godfrey observed that the duration which a furnace could be kept in-blast was typically only '20–50 days and [shutdown took] place in consequence of horse being formed in the furnace.'[31] A blast furnace's efficiency is directly related to the length of time it is kept in operation. By the 1870s, year-long campaigns were the norm, and continuous operation for a decade or more was not unheard of.

Throughout his trip, Godfrey recorded observations and made suggestions for 'the erecting of a new smelting works and improve[d] blast furnaces.'[32] It is obvious that Godfrey had undertaken this tour with precisely that goal in mind. His recommendations can be divided into four general categories: smelting, transportation and communications, better utilization of natural resources, and cost reduction. Possibly the greatest problem was, and remains to this day, Kamaishi's terrain and remote location. Without providing great detail, Godfrey suggested connecting the mines, ironworks, and harbor with roads, or a tramway, or both. He considered 'Ohashi and Sahinari [*sic*] . . . the best locations' providing 'a new road suitable for carts' be constructed between the ironworks and the harbor.[33] He also recommended that a reforestation program be undertaken and some type of plan for regulating the consumption of trees be adopted to combat the growing scarcity of wood for charcoal. As a potential way to remedy what appears to have been an impending fuel shortage, he also suggested the possibility of importing coal or coke from elsewhere in the country. 'Fuel might be brought in from the South and in return pig be taken back to be manufactured there into wrought iron.'[34] Moreover, Godfrey recommended the construction of larger (more efficient) furnaces closer to the sea, that is transportation by ship, and the use of gunpowder in mining. The last three items are also related to economizing operations.[35] Large blast furnaces are generally more efficient than smaller ones; and blasting a rock face is usually more efficient than attacking it with picks and hammers.

Most literature which discusses the design, construction, and subsequent supposed failure of Kamaishi Ironworks, focuses on an argument between Ōshima Takatō and Louis Bianchi, the foreign adviser hired by the Meiji government to oversee the mining and metallurgical related details of the Kamaishi project. Implied in these studies is the belief that Bianchi was responsible for the design of Kamaishi. However, Bianchi was not part of the original design plan; he was not even a government adviser at the time.[36] His later encounter with Ōshima was related only to finding a suitable location for the government's 'modern' ironworks.[37]

Another often repeated, but never substantiated, claim regarding the basic design of Kamaishi ironworks is that it was based on Yamao Yōzō's plan.[38] Based on a comparison with Godfrey's report, written between July and September 1872, and the proposal for an ironworks at Kamaishi submitted by Yamao and Itō Hirobumi to the Council of State (*Dajōkan*) and Ministry of Finance on 15 February 1874, it would not be unreasonable to state that Kamaishi was Yamao's plan based on Godfrey's initial observations and suggestions.[39] The common assertion that this plan was Yamao's idea alone is unsupportable.

Based on the idea that Japan was rich in iron and coal, and that Japan should not be reliant on foreign imports for the basic materials of 'civilization,' Yamao's and Itō's proposal argued that an ironworks should be established in Heigori, at Kamaishi, so that Japan could produce the materials for its own railroads, telegraph system, and ships, as well as supply its own military iron needs. Beyond this fundamental statement of import substitution and 'civilization' building is a fairly detailed budget proposal which outlined plans for facilities at Kamaishi and Nagasaki. The plans include a railroad from the mines to Kamaishi harbor and three modern blast furnaces deemed capable of producing 12,000 tons (10,886.4 tonnes) of pig iron per year. There would also be a refinery at Nagasaki, with 12 furnaces, where the pig iron from Kamaishi would be shipped for further processing. The total cost for the venture was estimated at ¥830,000.[40]

Except for the commentary on import substitution and nation building, Yamao's proposal is remarkably similar to Godfrey's recommendations. He proposed constructing larger, more efficient furnaces, connecting the mines, furnaces, and harbor with a railroad, and suggested the possibility of shipping pig iron produced in the Heigori region on the return trip of ships that had brought coal or coke to Kamaishi from the south. It is no coincidence that Nagasaki was the southern destination of the public works ministry proposal, being the location of some of Japan's best known coal mines at the time. The ministry's proposal could even be considered more progressive than what Godfrey had suggested. Godfrey only mentioned 'roads suitable for carts' or possibly a tramway. Yamao and Itō moved to infuse their new facility with 'civilization' by directly proposing a railroad.[41]

On 21 May 1874, the Meiji government established the Kamaishi branch office of the Ministry of Public Works' Mining Department. As part of the process, the Hashino, Ōhashi, Sahinai, and Kuribayashi mines were nationalized; and Ōshima Takatō and another technologist, Koma Rinosuke, were appointed as Kamaishi's managers.[42] According to some sources, there was a magnificent ground-breaking ceremony for the government's new iron works on 10 August 1874; actual construction began in January of the following year.[43] Interestingly, however, at the time there was mention neither in the Japanese- nor English-language press of any ground-breaking ceremony at Kamaishi – or even statements that the government was planning to construct an ironworks until years later.[44] This is in stark contrast to the publicity received by the Tomioka Silk Filature at nearly every stage of its construction. Even the opening ceremony for a *noile* (waste silk spinning) facility in Gunma prefecture drew more official and press coverage than Kamaishi.[45]

The man often portrayed as the villain of Kamaishi is Louis Bianchi, the Meiji government's foreign adviser for mining and metallurgy at the ironworks. His notoriety stems from a number of factors, including having the government choose his proposal over Ōshima's – father of Japan's modern iron industry – regarding the location of Kamaishi's furnaces. It is also typically implied that Kamaishi's final design was based on the same proposal, and that Kamaishi's so-called failure was the result of Bianchi's lack of familiarity with Japan.[46] Moreover, it is assumed that Bianchi abandoned the project, or at least implied that he quit *after* getting his way. By all accounts, Bianchi had a rather abrasive personality and was known to be quite stubborn.[47] That he argued with many of the engineers and managers – Japanese and foreigner alike – does little to help his reputation. And all of this has been shrouded in mystery because no one has bothered to find out anything about him.

Born in 1836 into a family which owned an ironworks in Hockeroda, in the Principality of Schwarzburg Rudolstadt, Louis Bianchi enrolled in the Freiberg Mining Academy in October 1856, after previously studying at a Klaustal mining school.[48] Studying at Freiberg until 1859, Bianchi completed course work in analytical chemistry, general mine surveying, general and iron metallurgy, and welding.[49] Throughout his academic career, Bianchi remained in good standing at the academy. He was a member of the Franconia Corps, a fraternal organization,[50] and, with the exception of a 'very minor legal matter' involving a local tailor, his time in Freiberg appears to have been rather unexceptional.[51]

Coming to Japan in early 1874, Bianchi signed his contract with the Meiji government in March of the same year. It is possible that there was a connection between Bianchi and Godfrey, that they knew each other, or at least knew of each other prior to Bianchi's arrival in Japan. Godfrey also attended the Freiberg Mining Academy, during 1860 and 1861, and took a number of the same classes with the same instructors.[52] At the very least, the two men knew each other in Japan as Godfrey was Bianchi's superior and a signatory on his contract.[53]

Within a month of signing his contract with the Meiji government, Bianchi was in the Heigori district with Yamao Yōzō surveying the region, in part, to see if there was any coal near Kamaishi that could be used at the ironworks. The search must have been difficult and one can only imagine Bianchi's frustration at operating within the greater constraints placed on the Mining Department. Unlike the more 'enlightened' departments, like railroad and telegraph, mining did not even have the proper equipment with which to carry out the survey. In a letter to Itō, Yamao rationalized their request for the boring equipment with which to take core samples of the rock outcroppings. He noted that the Railroad Department had the required equipment, but that borrowing it required too much time. If the Mining Department had its own boring set, he argued, not only could it be used to check for coal at the Kamaishi and Kosaka mines, the department could also share it with the Manufacturing Department.[54] It somehow seems ironic that the Railroad Department had the tools with which to locate coal, yet the department charged with mining this mineral fuel did not!

During the late spring and early summer of 1874, Bianchi continued to survey the region with Ōshima Takatō.[55] The two men studied the Kamaishi area in an effort to find the best location for the government's new ironworks. By all accounts, Ōshima and Bianchi could not come to an agreement and fought bitterly.[56] Both men are known to have been stubborn. Ōshima's reputation, in fact, had earned him the nickname *inoshishi*, or wild boar: one who stubbornly rushes headlong into danger without considering the consequences.[57] Ōshima must have resented having to translate Bianchi's proposal, one with which he disagreed, for presentation to the government, but Bianchi was Ōshima's supervisor and he had little choice. In an unprecedented move which violated the chain of authority, however, Ōshima also submitted his own ideas on the subject to the government in a parallel proposal.

An examination of the two proposals reveals that their only similarity is the handwriting in which they were written. Bianchi proposed a site at a location named Suzuko. Selection criteria were evaluated based on the site's proximity to transportation, access to water, proximity to ore and fuel, available land, and cost. Ōshima's proposal, *Otadagoesetsu*, named for the site Otadagoe, was based on building the ironworks in a location that was sheltered from the elements:

The topography of *Otadagoe* to the east makes it the best choice. It is surrounded by mountains on three sides: north, west, and east, only the south is open. Throughout the four seasons there are no rainstorms, and even in severe winters, the cold will not stop production day or night.[58]

Every other aspect of Ōshima's proposal was subverted to fit his choice of a naturally sheltered location. Because the valley was too small to accommodate an ironworks, he suggested flattening the mountain tops to gain the required storage space. He even proposed running iron pipes around the base of the mountains for a distance of 9.6 miles (3.9 kilometers) to bring water to the location. (See Appendix I for complete text of the proposals).

Ōshima's proposal violated every convention of nineteenth-century ironworks design. Yet he had successfully built over a dozen blast furnaces which were responsible for nearly all of Japan's Bakumatsu and early Meiji era iron production in Western-style furnaces. Some scholars have even identified the factors he seemingly ignored in his Kamaishi proposal as significant in the design and location of the furnace sites at Hashino, Ōhashi, and Sahinai.[59] What then, led Ōshima to submit a proposal which seemingly disregarded the most basic principles of ironworks design? Ōshima was motivated by a number of factors including experience, principally his Mito reverberatory furnace having been damaged by a storm, and the understanding that his Ōhashi furnaces would have to be demolished to build a new ironworks at the site recommended by Bianchi.[60] Ōshima was quite familiar with the Kamaishi area and its climatic peculiarities. After all, he had spent over a dozen years in the area building blast furnaces in as many locations. While not typically a concern of 'modern' ironworks design, Kamaishi's weather would prove, if nothing else, a hindrance to the efficient construction of the facility.[61] Whether Ōshima was correct in his assertions, however, was irrelevant. Victorian era 'progress' was frequently identified as man's technological triumph over nature. Any suggestions to the contrary were 'traditional' and 'backwards' and unacceptable from the government's perspective.

There were other possible problems with an Ōshima-designed ironworks as well. Yamao's proposal to the government mentioned three blast furnaces with approximate output capacities of 11 tons per day (12,000 tons/10,886.4 tonnes per year). All of Ōshima's Kamaishi area furnaces produced approximately 500 *kanme*, approximately 2 tons (1.81 tonnes), of pig iron per day. At the time of Godfrey's report (September 1872), the total output of Ōshima's Heigori region furnaces was approximately some 200,000 *kanme*, or roughly 825 tons of pig iron per year.[62] It is clear that the Japanese government intended to build an ironworks which utilized large, modern Western-style blast furnaces. No one other than possibly Ōshima intended to build a new ironworks at Kamaishi that relied on the region's extant technologies.

While Ōshima never stated what type of furnaces should be utilized, the assumption is that they would be similar to those he had already built in the area. Ōshima mentioned five furnaces in his proposal, coincidently the same number as were standing at Ōhashi, but much past that, he gave no direct indication as to style or design. He recommended a horse-drawn trolley system to supply five furnaces with iron ore. Based on the daily consumption rate of Ōshima's furnaces at Hashino and Bunkozan, five furnaces would have required more than 48 tons (43.55 tonnes) of ore and fuel per day, an amount within the capabilities of the horses he suggested. If, however, he was thinking of furnaces along the lines of what Yamao had proposed, requiring over 200 tons (181.4 tonnes) of iron ore per day, not to mention flux, the horses would not have fared as well.[63] In fact, well before the Meiji government even considered building its new ironworks, British engineers were debating the economic and functional utility of horse-drawn carts for mine haulage.[64]

Some have argued that Bianchi's proposal was accepted because he was a foreigner and that the government had a policy of 'blind respect' for foreign employees.[65] Others have simply stated that the government had 'greater respect for the opinions of foreigners.'[66] As demonstrated in previous chapters, nothing could be further from the truth.[67] And much like the circumstances surrounding the government's choice of technique at Tomioka, there does not appear to have been a careful technical examination of the proposals for Kamaishi. There was no discussion of the problems with Ōshima's furnaces' linings, and no inquiry into why the furnaces could only be kept in-blast for short periods of time. In fact, no one ever solicited a proposal from Ōshima in the first place! Similarly, however, there was no evaluation of which Western smelting technology was appropriate, whether Japanese ironworkers could operate large complex furnaces, or if there were viable alternatives. It was simply assumed that Kamaishi would rely on the latest Western technologies.

The government wanted to build a 'modern' ironworks to supply all of Japan's 'nation building' iron needs. There was, however, more to it. What the government was trying to build in the mountains of north-eastern Japan was their own version of Krupp. One foreign observer at the time even stated that the Meiji government intended to make Kamaishi into another Essen, 'with or without the help' of that world-famous manufacturer.[68] Yamao wanted a large ironworks from the start, one that would supply all of Japan's domestic and military iron needs, and this was in line with his other activities at the public works ministry.[69] All of his projects – building a railroad, lighthouses, and a telegraph system – were concerned with raising Japan's level of 'civilization' in concert with Western ideals. 'Civilizations' were measured in miles of railroad track, through mechanized industry, and 'permanent' structures.[70] Meiji leaders endeavored to quickly assimilate these elements into Japan's industrial and urban landscapes. All of Yamao's projects required the use of iron, and they demanded it in quantity.[71]

From the perspectives of either importing 'civilization' or technical rationality, Ōshima's furnaces were unacceptable. If one sought 'modern,' these artifacts from the West were obsolete. They were old even at the time Ōshima was first translating Huguenin's book. By the time Ōshima and Yoshii toured the Kamaishi region in 1873, Ōshima's furnaces – dubbed Japanese- or Ōshima-style in Japan – were referred to as 'old-type' in the West. This fact would not have escaped Yoshii Tōru, the author of *Kōgyō yōsetsu* (*The Theory of Western Mining and Metallurgy*), a translation and compilation of four British texts on mining and metallurgy printed between 1873 and 1876, which identified the style of furnace used in the Kamaishi area as out of date and inefficient.[72]

To make matters worse, Ōshima's adaptations to the furnaces, usually credited to his genius, made them wholly unsuitable from an ideological perspective. The most 'modern' element of these furnaces was their cast-iron blowing engines. Huguenin's original plans called for a double vertical piston bellows. For reasons unknown, perhaps insufficient water power or a lack of suitable cast iron, Ōshima replaced them with a traditional Japanese box bellows.[73] Made from Japanese cedar with seals of racoon skin stuffed with rice hulls, although efficient, these bellows were 'traditional' and therefore unacceptable.[74]

Spurned by the government's decision, Ōshima nevertheless remained in his position serving faithfully at Kamaishi. He and Bianchi continued to tour the area in an effort to locate suitable sources of coal for the ironworks and other public works ministry projects. Ōshima was able to assume greater authority in the Mining Department after Yoshii retired in April 1875.[75] However, Bianchi was still directly responsible for Kamaishi, and most of Ōshima's efforts after his promotion were directed at the larger aim of developing Japan's mining capacity. Often his tasks were administrative in nature. By the end of October, Ōshima was sent to the Ikuno and Kosaka silver mines and the Sado gold mine where he would serve out his days in the Ministry of Public Works as Executive Mining Director.[76]

As if an omen of things to come, Bianchi's return to Kamaishi from Tokyo in May 1875 would be problematic; his ship was forced to wait at anchor in a natural rock harbor for two days as a storm raged along Japan's northeast coast.[77] Frustrated at having been delayed, Bianchi immediately set about the task of trying to build the government's new ironworks. He expected to be intimately involved in the process, so much so that even his residence was to be constructed within the ironworks' compound for the sake of efficiency. If nothing else, however, Bianchi's time in Japan appears to have been one source of frustration after another. First he was faced with a lack of equipment when surveying the area with Yamao, he then became embroiled in an argument with his assistant, Ōshima. His return to Kamaishi was delayed by the weather, and he then faced numerous other obstacles – technical, administrative, and personal – once construction began.

Much like the dilemmas faced by Tomioka's technologists, one of the greatest tasks facing Bianchi was finding or making the basic materials from which to build the ironworks. He needed to make bricks suitable for building construction and firebricks capable of withstanding the rigors of iron smelting in a modern blast furnace. Estimating that they would need nearly half a million bricks and over six months just to make them, Bianchi was anxious to get to work because only then could construction begin. Stone for the furnaces was available on-site and Bianchi recommended that workers begin to collect it immediately. By the end of the month, after a number of failed attempts, Bianchi reported that it would not be possible to manufacture satisfactory firebricks at Kamaishi. Even clay brought in from the Izu region was not suited to the task because of a high quartz content. Other clays had too much sulphur, which is detrimental to quality iron production.[78]

Bianchi recommended that the government order all boilers, machinery, and iron furnace parts from England. In the end, even the majority of Kamaishi's firebricks would be imported from that country. All of the iron parts for Kamaishi's blast furnaces, as well as refractory materials and furnace appliances, for example, boilers and blast engines, would be purchased from various British firms.[79] By the summer of 1875, Kamaishi's workforce was also largely populated by British mining and civil engineers, and by British mine and railway workers. According to at least one of these engineers, Gervaise Purcell, this too, appears to have been a source of frustration for Bianchi.[80]

Traveling on-board the steamer *Colorado* from San Francisco, Purcell arrived in Japan on 14 March 1874.[81] Originally hired by the Ministry of Public Works' Railway Department, he transferred to the Mining Department in May of the following year and came to Kamaishi a month later with the task of constructing a railroad linking Kamaishi, Ōhashi mines, and the harbor. His first months at Kamaishi were filled with settling into the rather rustic environs, working on plans for the new railroad, and surveying. He seems to have adapted well, uncomfortable housing conditions notwithstanding, and always had time for the occasional pheasant hunt. Nevertheless, Purcell and Bianchi argued a number of times during the former engineer's first two months on the job. Bianchi even went to the trouble of sending Ōshima an official letter in August 1875, stating that he did not want any more Englishmen sent to Kamaishi, and that he wanted to work only with fellow Germans. According to Purcell's comments on the incident, 'he [Bianchi] and I do not agree, as he says, and apparently intimat[ed] that it obstructed the work.'[82]

For all his efforts, Bianchi was not the person responsible for designing Kamaishi; his job was to oversee construction. The design of the furnaces and related equipment, based on the public works ministry's demands, was the project of David Forbes, another British engineer.[83] Forbes, however, never actually visited Kamaishi, remaining in London while he served as consulting mining engineer to the Meiji government.[84] His task was to design the smelting works and procure and send the necessary equipment to Japan from England.[85]

Born on the Isle of Man in 1828, Forbes excelled in his studies at the Athole Academy. Showing an affinity for chemistry, he left the Isle of Man and attended an academy at Brentwood, Essex. In October 1844, he transferred to the University of Edinburgh to study under Dr George Wilson, a famous chemist.[86] At the age of 16, his talents brought him an appointment to the position of assistant chemist at the university. For most of 1846, he worked in the metallurgical laboratories of John Percy, a Queen's Hospital physician with an interest in geology, who had developed new methods for extracting silver from its ore 'to supplement his meager salary.'[87] Forbes' first practical experience as a metallurgical engineer came while he was employed by Evans & Haskins, a Birmingham nickel smelter, at what was to become the company's nickel and cobalt refinery in Espedal, Norway.[88] There he quickly took charge of constructing the company's mining and smelting works in the Norwegian Alps.

Having been made a partner in the firm, Forbes left Norway in 1856, and traveled to South America where he largely remained for the next six years. Searching the continent for exploitable sources of nickel and cobalt for Evans & Haskins, Forbes pursued the study of mineralogy in Panama, Chile, Bolivia, and Argentina. A prolific writer, Forbes published 58 papers, many the product of his extensive knowledge of South American mineralogy and geology.[89] Forbes was particularly interested in the effect of high temperatures on minerals. He spent time studying volcanic materials and 'took advantage of the metallurgical operations' at Evans & Haskins where he exposed various rocks and minerals to the extreme temperatures and pressures of the blast furnace.[90]

Forbes had a sense of adventure and frequently became embroiled in local politics. In 1848, for example, he received personal thanks from the King of Norway for his efforts in putting down a revolutionary movement. Forbes had armed 400 of the smelting facility's workers and fought on the government's side.[91] Decades later in Bolivia, he was again involved in a revolution. This time he did not fair as well and was bayoneted twice for his troubles.[92] Leaving South America, Forbes traveled to the United States, Egypt, Spain, Germany, and again Norway where he studied the mineral deposits and mining methods of the various countries. During the last dozen years of his life, from 1864 until 1876, he worked as a consulting mining engineer. It was in this capacity that he served the Meiji government. As Foreign Secretary of the Iron and Steel Institute, from 1871 until his death, Forbes dedicated himself to studying and reporting the 'details and progress of the iron and steel industries' outside of England.[93]

Ordering the furnace components from the Teesdale Iron Works, Stockton-on-Tees, Forbes designed two blast furnaces that sported 'all the most modern improvements.'[94] He incorporated hot-blast stoves and a newly patented hydraulic mechanism for regulating the collection of the waste gases that were used to provide heat to the stoves and the boilers for Kamaishi's steam powered blowing engines. Each furnace was configured to produce approximately 75 to 80 tons (68 to 72.58 tonnes) of charcoal pig iron per week.[95] As clearly seen from Forbes' description of the furnaces and components, the Meiji government's new ironworks at Kamaishi was to be large, efficient, and above all – 'modern.'

Even as Forbes was involved with the details of designing Kamaishi, the government's plans were evolving. From 1874 through 1876 the number of proposed blast furnaces changed. Reports range from two to four, and furnace outputs similarly varied from 10 or 11 tons (9.07 or 9.99 tonnes) per day up to 25 tons (22.68 tonnes).[96] The issue of fuel was similarly unsettled: some reports stated that Kamaishi would rely on charcoal, some reports claimed coal, others both.[97] On at least two occasions, and as late as mid-1878, Yamao was at Kamaishi trying to help settle the fuel issue. The problem faced by the government was that if Kamaishi's furnaces proved to be as efficient at converting the region's iron ore into pig iron as planned, in all likelihood there would be a shortage of charcoal.[98] Also revealing of the government's evolving plans for Kamaishi, the idea of shipping pig iron to Nagasaki for refining was abandoned by the time the furnaces' metalwork and machinery were shipped to Japan during the autumn of 1875.[99] The new ironworks at Kamaishi had also become a refinery and mill. Forbes accordingly ordered puddling and reheating furnaces, a forge train, machinery for milling iron plate, rail, and bar, steam hammers, shears, and saws for hot iron, a roll lathe and crane. In short, according to Forbes, 'all the necessary appliances of the most *modern* construction.'[100]

Purcell too, found himself designing and building a railroad whose configuration was often subject to change. At the time he began working on the railroad from Ōhashi mine to Kamaishi harbor, for example, although receiving a hint that he should plan on structures that could support the weight of a 20-ton (18.14 tonnes) locomotive, no decision had been made whether the railway would use horses or the steam-powered alternative.[101] Purcell did not seem to mind the additional challenge of building to ever-changing specifications. He simply opted for the more heavy-duty alternative and built accordingly.

The ad hoc nature of the government's project, seemingly not a problem for the British engineers at Kamaishi, proved almost debilitating for Bianchi. With apparently much free time at their disposal while the government decided the particulars of the ironworks, Purcell and his companions frequently toured Kamaishi village and the surrounding countryside taking in its breathtaking sights and the local culture.[102] Bianchi, however, appears to have never joined his colleagues on their excursions and seems to have spent much of his time fighting with Koma Rinosuke over any number of issues. One in particular, deciding on office and mine regulations, was especially irksome. Demonstrating his mounting impatience and irritation with the Meiji government, in his official report on Kamaishi, Bianchi demanded that all the details be settled immediately. Finding suitable materials from which to make bricks was similarly taking too much time and money. According to Bianchi, erecting the new ironworks at Kamaishi was a very large undertaking and things were simply taking too long.[103]

When Bianchi's contract expired in March 1877, he had no problem with leaving his position. His time as the Chief Mining and Metallurgical Engineer for the Meiji government was nothing but frustration. He did not like the working conditions and apparently did not like the people with whom he worked. His departure had little effect on the project and did not even warrant a mention in Purcell's diary.[104] But Bianchi was not the irresponsible foreign adviser who abandoned the project after getting his way in the dispute with Ōshima, as is often the implied by his 'disappearance.'[105] Itō Hirobumi seems to have considered the problems between Bianchi and Ōshima unfortunate and a problem which contributed to Kamaishi's early demise. However, he criticized neither engineer for their inability to get along.[106] Neither villain nor opportunist, Bianchi was simply one of Meiji Japan's many foreign employees who largely vanished into the discourse of Meiji industrialization.[107]

After Bianchi's departure, construction proceeded despite numerous obstacles. The Kamaishi area was racked by earthquakes. When the ground was not shaking, there was frequent flooding. On numerous occasions, sections of Purcell's railway were washed away by raging torrents created by the heavy rains in the mountains.[108] It appears that Ōshima's concerns over the weather may have been valid, and other Japanese engineers at Kamaishi seem to have been concerned with other potential location-related problems. Upon final testing of an experimental friction brake on a steep 'one in seven grade,' Mori Jusuke,[109] the manager who replaced Koma Rinosuke at the ironworks, complained to Purcell that the 'furnace had been put in a very bad place!!!'[110] Whether or not Mori's comments were related to the area's topography is difficult to ascertain. Purcell and the other British engineers appeared unconcerned over the steep and rugged terrain that surrounded the ironworks. Bianchi, for his part, other than making note of the conditions and urging caution, especially near Shindenmae village, the area adjacent to the Kamaishi and Ōhashi mines, seems not to have considered the terrain problematic.[111]

Furnace construction was the job of William H. B. Casley, a British ironworks manager from Stockton-on-Tees, who had arrived at Kamaishi with the furnace components in early 1876.[112] In keeping with official government practice, a Japanese manager, Yamada Jun'an, was also responsible for daily operations. Yamada was trained in metallurgy at the London Mining College and later served at the Bleavanon Mines in South Wales. Upon his return to Japan he was appointed to serve at Kamaishi. The two men, along with Mori Jusuke, spent much of 1877 erecting Kamaishi's two 59-foot (18 m) tall blast furnaces and installing the furnace appliances. Things appeared to be going well, but in early 1878, toward the end of his second contract with the Meiji government, Casley wrote to Yamao Yōzō, then minister of the Ministry of Public Works, requesting permission to break his contract.[113] We do not know why Casley wanted to leave Kamaishi; however, Purcell provided him letters of introduction to his friends Wheeler Dunham and Matthew D. Mann, a prominent New York physician, just before his departure.[114] On the evening of 11 January 1878, Mori threw a sumptuous farewell banquet in Casley's honor. The handful of guests ate, drank, and sang to the *shamisen* into the early morning hours.[115] Casley left Kamaishi the following morning and a few weeks later boarded the British steamer *Oceanic* bound for London via San Francisco.[116] His departure left Yamada to supervise the remaining British ironworkers, oversee construction of Kamaishi's blast furnaces and appliances, and blow-in the furnaces.

Yamao Yōzō was back at Kamaishi in the spring of 1878 to inspect progress on the facility. Perhaps Casley's departure gave him cause for concern. Upon arrival, he learned that construction of the railroad was proceeding well, but that the charcoal issue still needed to be settled. He instructed Yamada to see to it that the problem was solved in an expedient manner.[117] For the next two years, Yamada put all his talents to the test ensuring that the government's efforts to modernize the iron industry would not be spent in vain. Despite setbacks, such as an early morning fire which destroyed a number of buildings around the office in late October 1879, Yamada dedicated his utmost attention to the project.[118] For his part, Yamao arranged for additional funding and appropriations of land in the area surrounding Kamaishi. The government even appropriated land on which to begin a reforestation project to insure a continuous supply of charcoal at the ironworks.[119] Under constant pressure from the Meiji government, especially from Yamao Yōzō, Yamada saw to the general completion of the facility by the following year.[120]

With most of Kamaishi's equipment in place, one furnace was blown-in on 10 September 1880. Three days later, at 10:30 a.m., the iron began to flow.[121] Although both furnaces were ready to be put into service, the plan was to rely on one furnace and hold back the second as auxiliary. Perhaps this idea harkened back to the days of Ōshima's Hashino and Ōhashi furnaces when 'round the clock operation relied on switching between two or three furnaces. Initial production at Kamaishi stood at approximately 7 tons (6.35 tonnes) of pig iron per day but increased steadily with time. Because fuel consumption was high, however, over 10,000 *kanme* (approximately 35 tons/31.75 tonnes) per day, the plan was to improve furnace efficiency. The forests surrounding Katsushi and the other Heigori villages where wood was cut for Kamaishi's charcoal would be rapidly depleted if improvements were not made.[122] The Kogawa charcoal-producing facility on which Kamaishi relied was similarly unable to handle the demand. Plans were quickly drawn up to utilize coal and surveyors were sent to the Kuji coal pits in Aomori prefecture to inspect the quality and supply of coal at that location.[123] Kuji's coal seam, however, was thin and the quality unsuited to iron smelting. Other sources had to be found.

While government officials and Kamaishi's engineers scrambled to find alternative sources of fuel, disaster stuck at the Kogawa charcoal facility. On 9 December 1880, less than three months after beginning production, a fire destroyed the facility, 15 other buildings, and the majority of Kamaishi's fuel reserves. The shortage of charcoal forced managers to shut-down operations on 15 December. Possibly a contributing factor to the fire, Kamaishi's managers quickly improved arrangements for storing charcoal at the facility. They were also able to extend Kamaishi's charcoal woods to 4,000 hectares and began making provisions to manufacture coke on site. By April of the following year, Yamao was back at Kamaishi to survey the damage. On the same trip, he also visited the Sado mines, perhaps to confer with Ōshima.[124] Yamao returned to Kamaishi again in June and July ostensibly for the same purpose.[125]

In an interesting move, one designed perhaps to combat growing dissatisfaction within certain government circles regarding the mounting expenses and lack of financial return at Kamaishi, Yamao proposed that the railroad built between Ōhashi and Kamaishi mines serve a dual purpose. He suggested that the narrow-gauge railway be opened to public use![126] Soon thereafter, in late October 1881, Hasegawa Yoshimichi, an inspector from the Ministry of Public Works, went to Kamaishi to further assess the situation. He appears to have been caught between rival factions at the ironworks. One group argued in favor of shutting-down operations until more fuel could be secured. The other group wanted to blow-in the furnaces with whatever fuel remained and run them for as long as possible. Perhaps swayed by the continuous efforts to improve fuel procurement, Hasegawa chose the second option.

Kamaishi's furnaces were restarted in March 1882, and furnace output increased 'without a hitch.'[127] During the month prior to resuming operations, Kamaishi's managers had brought 10,000 tons (9,072 tonnes) of coal to the ironworks and ordered the construction of 48 coke ovens. For a short time, at least, one furnace made the transition to coke. Based on Kuwabara Masa's observations, the furnaces were able to smelt with either charcoal, coal, or coke.[128] When using coal, however, Kamaishi's furnaces produced number 3 or 4 pig iron, not the same high-quality number 1 pig as had been the case when previously smelting with charcoal.[129]

Because the bituminous coal used at Kamaishi was highly friable, it became powder under its own weight – choking the furnace – as the charge descended.[130] When the switch was made to coke, little changed within the furnace and output steadily decreased during the 196 days in which it was in operation. Output was reduced because the furnace was chilling. At first the furnace started forming clinkers, partially smelted agglomerations of iron ore, fuel, slag, and flux. Later the charge fused into one solid mass – '*ittai katamari o nashi*' – and the furnace simply 'went-out.'[131] Shortly thereafter, the facility was declared a failure. Within a few months, the Meiji government abandoned its first entrepreneurial venture into the iron industry, selling Kamaishi Ironworks within the rubric of finance minister Matsukata Masayoshi's deflationary project.

Why not to build a modern ironworks?: the other foreign advice

If Kamaishi was nothing else, it was a blow to Japan's 'civilization building' prestige. After nearly a decade spent trying to modernize the iron industry, the Meiji government was left with little more than a few miles of salvaged railway materials, a clogged furnace, and a rather large slag heap. Despite foreign recommendations *not* to attempt such a grand project, Meiji leaders proceeded to build their ironworks, as was the case with the Tomioka Silk Filature, apparently throwing caution to the wind. Silk, however, was of great interest to the foreign powers and their involvement, welcomed or not, fueled government attempts to 'modernize' the industry. Great Britain was the world's largest trans-shipper of raw silk. The British had no capacity to produce this commodity on their own and thus tried to influence nearly every aspect of the Japanese silk trade to the best of their abilities. The French and Italians, although having had the productive capacity, were trying to reinvigorate their industries after suffering the ravages of the pébrine virus through the development of the silk and silkworm egg trade with Japan. In either case, Europe and the United States, at least for the time being, were dependant on shipments of raw silk and silkworm eggs from Japan and the rest of Asia. The same cannot be said for iron.

Foreign governments had little interest in having Japan developing its own iron industry. Official recognition – both Japanese and Western – of the government's efforts was correspondingly low. Contributing to the lack of foreign interest in this mineral resource was its availability and value in the international trade. Great Britain manufactured more iron than all other iron-producing nations combined.[132] Although their lead would be whittled away by Germany and the United States before the turn of the century, Britain supplied the bulk of the world's iron. An important trade commodity, approximately 60 percent of all British pig iron was produced for export.[133] From 1871 through 1877, a period roughly coinciding with the Japanese government's development of Kamaishi Ironworks and initial British interests in developing Japan's mining capacity, British iron exports averaged more than £28 million per year.[134] Perhaps frustrated by their inability to successfully develop an iron industry in India based on European methods, the British exercised more caution when it came to risking capital in Japan.[135]

Two factors contributed additionally to British hesitations: recent reports refuting previous claims of Japan's great mineral wealth; and the Japanese government's enactment of mining laws which excluded foreign capitalization of mining enterprises. First impressions of Japan's mineral wealth were promising. In fact, since Japan's earliest encounters with Europeans in the sixteenth century, the island nation had somehow managed to earn the reputation of being an Eldorado of sorts. When the British re-entered Japan in the mid-nineteenth century with the hopes of tapping into Japan's 'untold riches,' the stage was set for disappointment.

On the recommendation of Sir Harry Parkes during the summer of 1874, F. W. Plunkett, Secretary to the Legation, undertook to write a comprehensive evaluation of mines and mining in Japan. Completed the following spring, Plunkett's report, in no uncertain terms, cautioned against the investment of British capital in Japanese mines. 'Although metalliferous ores are found in so many parts of the country, it is extremely doubtful whether there are many of such a position, percentage, and character as would justify the investment of much capital in mining enterprise at present.'[136] Plunkett went on to state that the mineral wealth of Japan had been over-estimated, and that 'there are some good mines, but . . . they are the exception rather than the rule.'[137] If the Japanese government was to relax its mining regulations and admit foreigners and foreign capital to the industry, in Plunkett's estimation, it would be the Japanese who would benefit, not the unfortunate entrepreneur whose failure appeared preordained.

Sir Harry took Plunkett's warnings to heart and immediately dispatched a number of memoranda summarizing the report. He repeated that Japan's mines were not as rich as was previously believed. Moreover, Parkes leveled a number of criticisms at the Japanese government's restrictive mining laws which prevented foreign participation. He neither understood how the Meiji government could have been so short-sighted, nor saw any potential for legislative change in the immediate future. In short, even if the British government or its citizens could invest in Japan's mines, Parkes and Plunkett recommended against any official involvement.[138] Parkes went to the trouble of sending a copy of the report to the Ministry of Public Works, perhaps to soften Japan's legislative resolve. But the Japanese government, or at least Itō Hirobumi, already knew about 'the conditions' of mines and mining in Japan. Itō in fact had instructed J. G. H. Godfrey to supply Plunkett with the unpublished manuscripts upon which he based much of his report.[139]

Itō and other ministry officials had similarly received previous negative reports – ones which specifically questioned the technical and economic viability of Kamaishi Ironworks. Touring the Tōhoku region while employed by the public works ministry as part of their efforts to exploit Japan's oil reserves, Benjamin Smith Lyman had the opportunity to visit and report on the Kamaishi area. Much of what he wrote confirmed Godfrey's earlier observations regarding contemporary iron smelting practices. His comments, however, differed in two respects: his appraisal of the government's efforts to modernize the iron industry, and the area's resources. Noting projected necessary raw material requirements, 'the dearness of capital, the cheapness of labor, the greater cost of imported furnace materials, [and] the lack of workmen or superintendents familiar with the methods of large furnaces and of the latest blast heating apparatus,' Lyman argued that conditions in Japan were more 'favourable to the working of very small blast furnaces of improved native material, and probably still more to bloomery or Catalan forges' which were still 'common in many mountain regions of Europe and America.'[140]

Looking at the economics of the government's venture, Lyman estimated that a capital outlay of over ¥800,000 to produce approximately 5,000 tons (4,536 tonnes) of pig iron annually, meant that Kamaishi's pig iron would sell for approximately ¥30 per ton exclusive of labor or management expenses, and facility depreciation costs.[141] Lyman was largely correct in his calculations. In 1882, Kuwabara Masa reported that the cost of Kamaishi's pig iron, inclusive of labor and transportation fees, was ¥35 per ton, ¥22.85 per ton just for the raw materials.[142] Prior to the government's project, in 1874, pig iron was produced in the Kamaishi area for roughly one-third of that amount, approximately ¥9 per ton.[143] And Godfrey had erroneously estimated that the government could produce its iron for that price too.[144] Importing iron from Britain was considered the economical alternative, selling for less than what it cost to produce it at Kamaishi. In 1874 British iron would fetch between ¥22 and ¥25 per ton at the port.[145]

From the beginning, controversy also surrounded estimates of the area's iron ore reserves. In his report on Kamaishi mines, Kuwabara noted that three of the government's foreign advisers, Bianchi, Lyman, and Godfrey, were in disagreement about how much iron ore was at Ōhashi.[146] In a paper based on his earlier experiences in Japan read before the Geological Society of London in March, 1878, Godfrey stated that there were massive deposits of magnetic iron ore in the Rikuchū area. He estimated the deposits to be 50 to 200 feet (15 to 61 metres) thick and to stretch approximately 12 English miles (19 km).[147] Lyman's survey results indicated that there was far less ore than Godfrey had estimated, a possible total of only 140,000 tons (127,008 tonnes).[148] Bianchi's report was more vague, simply stating that after conducting test digs with Koma at Ōhashi, he found the majority of the mine to be layers of hard rock.[149] Yamao and Itō were committed to the project, however, and construction proceeded nonetheless. In the end, Godfrey would be proven largely correct, but his vindication would only come years after the government had sold Kamaishi.[150]

The generally negative assessment of Japan's potential for large-scale iron production continued throughout the period of Kamaishi's construction. Hired originally as an adviser at the Kosaka silver mines, then professor of mining and metallurgy at the Imperial University of Tokyo, Curt Netto also published a report which was quite critical of the government's efforts to modernize the iron industry. His greatest concerns appear to have been economic. Imported iron was less expensive than what the government could produce at Kamaishi. Efficient production would be hampered by the cost of charcoal, distance between the ironworks and suitable coal mines, 'difficult and underdeveloped' communications and transportation networks, and 'a circumstantial administration.'[151] Netto wondered if, given his estimates for Japan's future iron demands, the entire project was worth the effort.

Lacking the vision of 'civilization' possessed by the Meiji government, Netto discussed the advantages of remaining with the status quo. He argued that because of the dismal condition of Japan's roads, there was no advantage to building a railroad. The benefits of cost reduction would not materialize until goods were brought to a central location for trans-shipment by means other than a man's back. Following this logic, Netto also argued that it would be wasteful for the government to proceed with its plans to build more iron bridges throughout the country. If the government followed this advice, Netto reasoned, there would be a correspondingly lower demand for iron. He also claimed that Japan had a good coastal shipping network and the country's primary export goods, that is silk and tea, were lightweight and warranted neither a railroad, nor the expense of building one. Whereas Netto saw a possible future for shipbuilding, he stated that imported iron would be able to serve that need.[152] In sum, he argued against any future expansion of the iron industry. Netto predicted that the three proposed blast furnaces, two at Kamaishi, one at Nakakosaka, would be enough to meet all of Japan's iron needs. In fact, he cast doubt on whether or not those facilities would even be necessary, given the great expense required to build them and the cheapness of imported British iron.[153]

The Meiji government was faced with a dilemma. Foreign observers regularly characterized Japanese mining as 'backward' and recommended 'modernizing' the industry. They also criticized Japan's mining laws as overly restrictive, and belittled the quantity and quality of Japan's mineral resources. However, these same observers also wanted to participate, reforming the Japanese mining industry on their own terms, at times to reap whatever possible benefits there were. Moreover, the government's advisers provided conflicting information regarding the form which 'modernization' should take. To complicate the issue further, the Meiji government was also in the throes of domestically capitalized 'civilization' building. This required a medium of foreign exchange – obtained by exploiting Japan's copper, silver, and gold reserves – and a material from which to build 'civilization' – iron. Kamaishi Ironworks was built first and foremost for import substitution. From the start, Itō and Yamao were clear: Japan's so-called program of industrial and military development should not rely on importing foreign iron.[154] The foreign powers, however, and Great Britain in particular, stood to benefit from Japan's continued importation of cheap British pig iron, rails, and plate. In fact at the time the Meiji government was contemplating and building Kamaishi, the British export iron industry was attempting to expand its market to compensate for the declining price of iron.[155] As such, the British had no interest in Kamaishi's success.

Kamaishi was not built as an exemplar of 'civilization' per se. However, if the Meiji government could successfully build a 'modern' ironworks, something far superior to the numerous Ōshima-style furnaces that dotted Japan's proto-industrial landscape, the operation would be indicative of a certain level of 'civilization.' Because the construction of Kamaishi was undertaken with limited foreign involvement, success would also be demonstrative of true 'progress' and provide evidence to counter the many criticisms leveled in the Yokohama port press regarding the superficiality of Japanese attempts to 'modernize.' Throughout the early years of Meiji, the port newspapers were filled with critical evaluations of Japan's 'progress.' Some writers believed that assimilation of Western technology was good, if the Japanese were truly capable, but that it was necessary for the

Japanese to also become 'enlightened' and not simply be imitators.[156] Others like Richard Henry Brunton, an adviser to the Ministry of Public Works, questioned the abilities of Japanese engineers based on the education many received at the government's new engineering college.[157] With statements such as 'children, learn to walk before you run,' many Western observers questioned the nature of Japanese 'progress' and the pace at which Japan intended to 'modernize.'[158]

Japanese observers were also critical of the government's methods of promoting the mining industries. One author reproduced over 200 years of mining statistics, albeit abridged, in an effort to demonstrate that the adoption of Western methods had actually hindered production. The same author continued his critique of government-sponsored industrialization by noting that even if things turned out to be profitable at the government's mines, there was no way of knowing because information was withheld from the public.[159] What this writer did not realize, however, was that in all likelihood, the Meiji government was not interested in sharing information regarding developments at Kamaishi until its success was assured. In fact, unlike the regular progress reports on telegraph and railroad construction, there were almost no announcements in the major Japanese or English language newspapers regarding Kamaishi Ironworks from the time of the ground-breaking ceremony in 1874 until late 1880, when readers could only then find two line-announcements about which bureaucrat was going to inspect a group of mines that happened to include Kamaishi.[160]

Following Godfrey's advice, Yamao Yōzō and Itō Hirobumi had promoted a risky and expensive plan through which Japan could become self-sufficient in iron production. They had disregarded the warnings of any number of experts, Japanese and foreign, as to the economic and technical viability of their facility. If they succeeded, their plan was brilliant. Failure, however, if it occurred, would not be theirs alone. It would be a blow to Japan's international prestige, the government's domestic authority, and a step backwards for 'civilization' building. As a result, there was a paucity of public information regarding the government's ironworks until late in its development.

Unlike the silk reeling industry, where Meiji leaders were able to import fully functional machines and steam engines, many of the design and much of the smelting processes in nineteenth-century iron production relied on trial and error. Based on mineralogical assessments of ore and fuel, metallurgical engineers followed certain fundamental principles when designing a blast furnace. With this information, the best advice from any number of metallurgical handbooks published as late as the turn of the century was to find another furnace which operated successfully under similar circumstances and copy its proportions.[161] Once in-blast, a furnace still required constant adjustments to compensate for variations in the fuel and ore. Iron smelting was not science. It was dirty and prone to frequent setback and failure. Although integral to building 'civilization,' iron smelting was imprecise and therefore, at the same time somehow antithetical to 'civilization.'[162]

Disposal of the Ironworks

When the Meiji government disposed of Kamaishi Ironworks in the summer of 1883, the decision was almost predetermined. After receiving an erroneous report from Itō Yajirō, an inspector from the Ministry of Public Works, stating that Kamaishi would run out of fuel and iron ore within a few years, the government quickly sold the ironworks for ¥12,600, approximately half of a percent of what they had invested on the project.[163] Selling Kamaishi was easy for the government. Not in the sense that it was a good deal for the buyer, though it was, but because it carried none of the ideological baggage of 'civilization' building. Unlike Tomioka Silk Filature, there was no prestige attached to the government's ironworks, no mystique, little public recognition, and only infrequent public mention. There was no debate about the sale. No minister or bureaucrat spoke on behalf of maintaining the ironworks for its prestige value as had happened numerous times when Tomioka came up for sale on Finance Minister Matsukata's auction block.

Even the prospect of being self-sufficient in iron production was not enough to persuade the government to keep Kamaishi. It would take another decade and the fear of war with China for this argument to be resurrected.[164] As with its founding, the demise of Kamaishi went all but unnoticed in the popular press. The only vestige of the government's grand plan that was not officially abandoned was Kamaishi's railroad. Before selling the facility to Tanaka Chōbei, an army supply merchant, government workers stripped Kamaishi of its only functional elements of 'civilization,' 15 miles (19 km) of railroad track and some of the steam-powered machinery, for use elsewhere in the country.[165]

Part of the problem for Kamaishi was an often contentious relationship between the Ministries of Public Works and Finance concerning efforts to 'raise Japan's level of civilization.'[166] While members of the finance ministry were interested in 'civilization' building, they wanted to keep the efforts within the ministry's budget. As Steven Ericson has demonstrated, jurisdictional arguments within the ministries over financing and building Japan's railroads interfered with department policy throughout the 1870s and 1880s.[167] The public works ministry only became interested in iron production after recognizing the correlation between iron, railroad construction, and 'civilization' building; thus, Kamaishi can be seen as an expensive extension of that ministry's railroad policy.[168] The new focus on iron smelting and import substitution after the formation of the Ministry of Public Works' Railway Department was part of the government's plan to build a railroad network in Japan. Yet if one wanted to stay on budget, it was far more economical to use imported British, and later American iron rails than it was to invest in an expensive, locally produced alternative.[169] Myopic as this may seem, the government had invested over two million yen in Kamaishi, yet only produced 5,812 tons (5,272.66 tonnes) of pig iron, and produced neither rail for railroads nor plate for shipbuilding. Although integral to railroad development, an ironworks was considered secondary to the visible evidence of 'progress and civilization.' With readily available supplies of good imported iron, having one's own iron industry at the time was considered an expensive luxury which the Ministry of Finance was unwilling to support.

Itō Hirobumi's retrospective appraisals of Kamaishi and the iron industry in general are demonstrative of the nature of the government's project. Returning to the notion that European 'progress' was illustrated by its 'iron weapons, iron warships, and railroads,' Itō lamented abandoning Kamaishi ironworks.[170] Although blaming the problems at Kamaishi on the iron ore and coal not being found within reasonable distances of each other, Itō continually reiterated that, when it came to producing its own iron icons of 'civilization,' Japan had made little or no progress until the late 1880s because of its failure to develop a viable iron industry.[171]

What Itō meant was that the *government* failed to establish a viable 'modern' iron industry. After a number of years of trial and error, Tanaka was successful at producing pig iron at Kamaishi in furnaces that Lyman would have called 'improved native material.'[172] Contemporary sources claim that Tanaka's iron, smelted in small furnaces similar to those designed by Ōshima four decades earlier, was of equal quality to, if not better than, Italy's famous Gregolini iron.[173] Hiring Komura Koroku to oversee operations, Tanaka put Kamaishi's British-made furnaces in-blast once again in 1893. His success caused quite a stir. As a result, the Meiji government launched a second investigation into its failure at Kamaishi Ironworks. Noro Kageyoshi, Professor of Metallurgy at the Imperial University of Tokyo, adviser to the Ministry of Agriculture and Commerce, and Komura's mentor, identified a number of problems with the former government facility. He stated that there was no shortage of natural resources, as Itō Yajirō had claimed years earlier. Rather, he focused on the difficulties of transportation and Japanese ironworkers' lack of practical experience in operating a large ironworks.[174] Noro's report implied that the government had given up too soon.

To foreigners in the treaty ports and the interior, the government's abandonment of Kamaishi Ironworks would be no surprise. For some in fact, it was a problem endemic to Japanese efforts at 'modernization' and industrialization. On numerous occasions foreign observers and advisers complained about this problem. Expressing his frustration with Japan's mining industry in general, Professor of Geography at the University of Bonn, Johannes J. Rein, for example, complained of the 'unsteadiness and constant desire for innovation on the part of the authorities, who could not patiently wait till the reforms begun should be carried through and tested.'[175] The British too, complained of the superficiality of Japanese 'progress' and 'modernization' in this regard. Richard Henry Brunton charged the Meiji government with having a spasmodic public works policy, which he likened to waves breaking on the shore: 'rushing with impetuosity' at one moment, 'followed by serene quietude.'[176]

Unknown to these observers, however, was the underlying nature of Meiji Japan's so-called industrialization policy. For the new imperial government, technological artifacts were multi-functional. Reeling frames and furnaces, railroads and brick buildings were imported based not on practical utility, but on their ability to represent Western ideals of 'civilization' in Japan. Japanese officials had long recognized the culture of materials, what I refer to as 'cultural materiality.' Bound to a progress ideology, officials argued publicly that iron bridges, cast iron machines, brick buildings, and railroads were necessary for strengthening the country's economy and military. These icons of the industrialized West would demonstrate

Japan's 'level of civilization.' Supporting these half truths were constant reminders that these artifacts of the West were permanent, precise, and 'modern.' Explicit was the understanding that this was the method by which one classified the 'civilized countries' of the West.[177] For Meiji 'modernizers' utility was not necessarily important. Men like Shibusawa Eiichi are known to have imported 'modern' silk testing equipment, only to have it collect dust in a warehouse because no one knew how to use it.[178]

'Civilization builders' spread Japan's resources thinly over any number of pet projects. When forced to operate within the constraints imposed by the finance ministry, however, it was relatively simple to abandon projects that showed little immediate promise, had little prestige value, or could be circumvented through the continued importation of cheaper foreign-made goods. Such was the case with Kamaishi. The government's ironworks was expensive, and it had no prestige value. No foreign power ever lavished praise on the Meiji government for its efforts to build a 'modern' ironworks, as was often the case with the Tomioka Silk Filature. In fact, much of the contemporary discourse was limited to speculation or criticism of the government's plans. Even the government's advisers cast doubt on the project and pointed out that it was less expensive to import British iron. Moreover, any government effort to reinvigorate Kamaishi was similarly confounded by changing Japanese attitudes toward the West – specifically the material artifacts of Western 'civilization.'

When Tomioka Silk Filature opened its doors for business in 1873, Japan was in the early throes of *bunmei kaika* materiality, that is the rampant adoption of things Western in an effort to become 'civilized and enlightened.' The same can be said of the time period when Itō and Yamao were proposing Kamaishi, and for the first few years of its construction. By the time Kamaishi was blown-in in 1880, however, the mood of unbridled absorption of Western material culture had begun to shift toward one whose appraisal was significantly more cautious and critical. Japanese commentators in fact decried the loss of traditional cultural attributes and the expense of such practices.

Gone from the popular press and much government correspondence was the language of *bunmei kaika*. Until the late 1870s, the discourse of Japanese modernization was punctuated with words such as 'civilization,' 'enlightenment,' and 'progress.' Any artifact of the West imported to Japan would similarly transfer these attributes. With time, this discourse changed. Lamenting the impending loss of Japanese culture and stimulated by the West's lack of recognition of Japan's 'progress,' the heterodox voice of Meiji thinkers arguing against the excessive acceptance of *materia Hesperia* was starting to be heard. What had once been support of any effort to adopt Western material culture was now criticism. Just as popular and official adoption of Western clothing, for example, was a topic of frequent cultural debate in the early 1870s, the issue was later revisited from the opposite perspective. Countering arguments against permitting government officials to wear *hakama* instead of swallow-tail coats while on official business as 'a retrograde step of the people from a state of civilization,' were commentaries that focused on the economics and practical aspects of abandoning Western formalwear.[179] Seemingly insignificant, the discourse on such superficial items as clothing, furniture, and eating utensils is representative of the greater issue.

Although the Meiji government did not promote traditional iron smelting techniques as the basis of import substitution, they could not ignore the facts. All of Japan's pig iron, over 30,000 tons (27,216 tonnes), had been smelted in either traditional Japanese *tatara* furnaces or Ōshima-type furnaces while the government's ironworks was under construction.[180] After nearly a decade of construction at a cost of more than ¥2,000,000, Kamaishi could boast less than one-fifth of that amount. After becoming finance minister in 1880, Matsukata Masayoshi turned a myopic eye towards any future potential at Kamaishi Ironworks and focused his ideological perspective on the economics of the issue.[181] A self-proclaimed proponent of laissez-faire policy, Matsukata advocated that 'industrial ventures be left to the private sector.'[182] His attitude toward the agricultural sector, however, including agro-industries such as silk reeling, and animal husbandry, is revealing of a less than consistent policy of funneling funds to favored developmental projects while denying others.[183] Matsukata was also biased towards supporting those areas of the economy that would increase Japan's export revenues.[184] As the government's industrial white elephant, Kamaishi fit none of the characteristics of the finance minister's privileged policies.

Originally promoted as part of Yamao Yōzō's plan to develop Japan's mining potential, Kamaishi was built in a storm of controversy. Almost every government adviser had recommended against a facility of such scale. Lyman argued that iron production would be inefficient and expensive. Netto stated that Japan did not even need an iron industry, given his forecast for a future of limited demand. Ōshima was similarly against the location and scale of the proposed facility. Since Kamaishi was based on his recommendations, only Godfrey appears to have accepted the government's plan, but he too, urged caution because of an apparent lack of fuel and poor transportation infrastructure. The British government also made sure that Meiji officials would know of their displeasure. Not being able to participate in the extraction of the nation's mineral wealth, Sir Harry Parkes notified the Ministry of Public Works of what he was sure would be impending failure. With evidence amassing against an independent, modern Japanese iron industry, Kamaishi remained out of public view.

For the Meiji government, Kamaishi was part of some plan to raise Japan's level of 'civilization.' Having already seen the great nations of the West in all their industrial glory while on the Iwakura Mission and at three international exhibitions, Meiji leaders relied on a progress ideology of materials in shaping their personal visions of Japan's future. In order to gain parity with the West, to be recognized as legitimate heir to Tokugawa political authority, Meiji Japan would be rebuilt in iron and brick. All of Japan's iron, the material from which to build 'civilization,' would be produced at Kamaishi. Projects of such importance, building a modern navy, railroads, and factories – the essence of the nation's 'progress' – could not be reliant on imported iron. Kamaishi would be Japan's Essen.

By the time the government sold Kamaishi Ironworks a lot had changed in Japan. Efforts to 'improve' Japan through the importation of the icons of 'civilization' had brought qualified approval from within and without. The new nation was faced with multiple crises including a failed civil war, cholera epidemics, a burgeoning popular rights movement, and a runaway inflationary economy. After failing to produce any significant quantity of pig iron at their new facility, Meiji officials decided to cut their losses – both to pocketbook and prestige. Kamaishi's 'failure' was not simply a technical or economic setback. By ignoring foreign advice, Meiji officials exposed themselves to a new commentary on the superficiality of Japanese 'progress' and attacks on the ability of Japan to 'advance.' Resigning themselves to an immediate future of reliance on cheaper imported iron, however, officials were able to continue building 'civilization' and do so at a lower cost. In the end, it would take another decade of internal political turmoil and a successful war with China before iron autonomy would again be a pressing issue.[185]

5 *Bunmei kaika to gijutsu*
Technology's role in 'civilization and enlightenment'

'A primer of civilization'[1]

In 1873, Matsumura Harusuke, a little-known writer, penned a thousand Chinese characters into a book which he called *Kaika senjimon, A Primer of Civilization*. He was surprised when a local publisher asked for permission to reproduce his work because he had only intended for it to serve as a guide to local school children and family members. Although not truly recognizing the potential significance of his work, Matsumura agreed and noted that his 'book contained words used enthusiastically in official notices and reports that should be put into ordinary usage.' While 'great men' were already familiar with these words used to describe 'civilization and enlightenment,' he noted, the primer would perhaps help future generations to understand the true meaning and spirit of 'civilization' and the government's pronouncements.[2] Alongside such words as 'Paris' and 'London' were the more ubiquitous phrases like railroad, mine, agricultural promotion, [machine-] reeled silk, brick, telegraph, civil engineering, steam, and, of course, machine.[3]

Matsumura may have been influenced by some of Fukuzawa Yukichi's early writings or by Nakamura Masanao's translation of Samuel Smile's *Self Help*. Matsumura included a number of words and phrases that reflected the popularization of such works and the power of individual self-determination.[4] His list also included some words that, while not new to Japanese society, reflected some of the darker sides of 'civilization' such as 'prostitute' and 'bribe.'[5] Much of his focus, however, remained on words and terms that exemplified many of the material aspects and institutions of *bunmei kaika*. By doing so, Matsumura unknowingly fell toward the externalists' side of the modernization debate. While many in Japan and Europe, the internalists, argued that the Japanese must first comprehend the *spirit* of Western civilization – liberalism and the spirit of scientific inquiry – others saw greater utility in the rapid absorption of Western material culture and the 'civilization' it represented.[6]

At issue is whether either party to the debate truly understood the externalists' behavior and motivations. While it is clear that there were times when internalist and externalist alike recognized that importing some aspect of Western material culture, railroads for example, would enhance the government's authority and prestige, in many instances the internalists may not have completely understood exactly why Meiji cultural materialists chose to transfer technologies that were uneconomical and extraordinarily difficult to disseminate. Most internalists, and even observers outside the debate, simply saw the rampant adoption of Western material culture as mimicry with no purpose.[7] Their quandary, however, strikes at the heart of the progress ideology of materials. Following beliefs in universal historical development based on man's mastery of nature and technical achievement, Meiji cultural materialists similarly recognized a hierarchy of materials. As man progressed from one stage of historical development to the next, so too did the material basis change to match his new level of 'civilization.'[8] Wood was part of the old order, iron and brick represented the new.

In making their own 'new world' the Meiji cultural materialists were able to take technology and technological artifacts and use them to create their own social and cultural frameworks. They were able to construct and reconstruct Japan's identity in terms of technological artifacts to match the country's rapidly changing socio-political and intellectual climates. Their 'civilization building' methodologies can be considered a battlefield of sorts. Intellectuals fought cultural materialists over the means for Japan's proper 'accession to the comity of nations.'[9] The intellectuals believed that Japan's international acceptance as a political equal depended on its acceptance of Western liberal ideals and institutions exemplified, for example, by the adoption of a national representative assembly. Meiji cultural materialists, however, as importers of the new technologies, were inventors in their own right. They reinvented technological artifacts in new situations and assigned their own social and cultural values along the way.[10] They were not necessarily concerned if anyone understood the 'spirit' of scientific inquiry or invention, only if their new artifacts would transform the environment in which they existed.[11] Cultural materialists did not take a holistic approach to 'civilization building': pieces of the West would construct a higher level of 'civilization' in Japan.

The cultural value of technological artifacts was firmly embedded in the rhetoric of 'civilization and enlightenment.' Regardless of focus, reform measures at all levels were couched in terms of 'civilization building.' Each artifact imported from the West would impart its socio-political values of 'progress' and 'civilization' once deposited in Japan. Writing in the *Tokyo nichinichi shinbun*, for example, a government propaganda organ of sorts, one commentator argued that Japan should give up the rickshaw in favor of horse-drawn omnibuses because they were the favored means of urban transportation in the 'great civilized and enlightened countries,' *daibunmei daikaika no kuni*, of England and France. Although many Japanese were afraid of horses, the author continued, people's fears would subside with time and the omnibus would bring a 'civilizing' influence to the cities of Japan.[12] Fukuzawa Yukichi and others often spoke of the railroad as a necessary element for the Japanese attainment of 'civilization' and international equality. In his *Bunkenron* published in 1876, and again four years later in *Minkan keizairoku*, Fukuzawa argued that the construction of railroads was an 'imminent necessity' if Japan were to equal the West.[13] In short, 'civilization and enlightenment' in Japan would be represented by Western material culture.[14]

Taking their cue from the West, Meiji thinkers and officials classified the nations of the world in a civilizational hierarchy based, in part, on technological achievement. According to these beliefs, Japan was 'semi-civilized,' superior to the 'primitive' countries that were helpless before the 'forces of nature,' but behind the 'civilized' nations of Europe and America that had conquered both the material and spiritual worlds.[15] All of this was relative, however, and a 'semi-civilized' nation such as Japan could be considered 'civilized' when compared to ones that were 'primitive.'[16] Men like Fukuzawa recognized that the West was not an ideal, it was not the end which Japan sought. Europe and America were rife with war, dishonest men and practices proliferated at every level of society. Even at this early date, the Japanese considered many Western city streets to be dangerous.[17] As such, the goal of 'civilization and enlightenment' was not to become Western per se, but was rather to surpass the West through the absorption and mastery of its material culture and some of its progressive values.

European and American civilizations had attained their dominant world positions through the technical achievements of the past half century, especially since the invention of the steam engine.[18] Many, including Ōkubo Toshimichi and Kume Kunitake, recognized the recent vintage of Western 'progress' and 'civilization.' It was beyond any doubt that Japan too, by internalizing the West's technological achievements, could occupy an international position similar to that of the European powers and America.[19] Fukuzawa held similar views. He argued that the West was only 20 or 30 – at the most 50 – years ahead of Japan in developing the latest technologies. Because of his belief in universal progress, he was certain that Japan one day would look back at an 'uncivilized' Europe![20] The esteem with which government official and commoner alike held artifacts of Western material culture helped construct and reconstruct Japan's new identity during the first decades of the Meiji era.

The first stage: Meiji techno-diplomacy

The first period of Meiji *techno-identity* formation, roughly 1868 to 1877, can generally be described as a time of near wholesale adoption/absorption of Western material culture in the name of 'civilization building.'[21] On an official level, government policies which promoted the adoption of *materia Hesperia*, from formal wear to railroads, should be considered acts which reflected the new government's sense of political urgency. Facing challenges from within and without, the Meiji government needed to construct an image that underscored it as the legitimate heir to Tokugawa political authority. Efforts to import and embed icons of the West were, on one level or another, Meiji foreign policy.[22] The physical manifestations of Meiji political identity were the icons of Western 'civilization.' As such, many of the government's actions during this period can be described as *techno-diplomacy*.

Meiji leaders hired foreign architects and engineers to design and construct ministerial buildings, of course built from stone and brick, to hasten their 'acceptance . . . and integration into a Europe-centered world order.'[23] In their efforts at promoting Japan's industrial base and the economy, officials insisted on importing cast iron machinery from Europe and the United States to bolster Japanese national prestige in a similar fashion. Rejecting viable, indigenously produced machinery that was typically labeled quaint or native, cultural materialists promoted the importation of iron machinery from Europe as exemplary of domestic and international prestige.[24] With regard to the construction and outfitting of filatures, for example, Hosokawa Junjirō, author of an early sericulture and reeling manual, and legislative scholar who was instrumental in founding the School of Western Studies (*Kaisei Gakkō*), forerunner of Tokyo Imperial University, argued that the best filatures are ones that relied on high-quality imported foreign (i.e., iron) machinery, housed in large, solid (i.e., brick) buildings.[25] Rather than consider a machine's or building's utility or expense, Meiji cultural materialists based their recommendations and choice of technique on a progress ideology of materials.

Concomitant with this belief, railroads and telegraph lines rapidly criss-crossed the country, while officials also promoted the importation and construction of iron-hulled steamships. Here, their aims were somewhat less unabashedly understood as promoting the government's authority. Itō Hirobumi and Ōkuma Shigenobu stressed that a publicly owned railroad would 'promote administrative consolidation' and thus bolster the new regime.[26] Sir Harry Parkes lent his support as well, pointing out to Meiji officials that a publicly-owned railroad would enhance the new 'government's authority.'[27] Interestingly, Itō and Ōkuma faced strong opposition from conservatives and military hardliners who failed to recognize the railroad's strategic value and resented any 'non-military' expenditures.[28] Ōkuma admittedly knew little about railroads and their construction. He did, however, recognize that railroads would promote national and government prestige. In an effort to persuade some of the more conservative members of the bureaucracy, he had Maejima Hisoka draw up estimates of expenditures and revenues, thus turning the issue into something at least partially economic.[29] Potential fiscal gains were enough to convince some of the conservatives who perhaps thought they could redirect any revenue toward building a *rich nation* and *strong army*.

On a more personal level, Meiji fashion was also decidedly Western. Pocket watches and Western-style umbrellas were all the rage. In 1873, ¥410,000 was spent importing the latter item alone.[30] Whereas fashion in the countryside remained distinctively Japanese, 'the hats and boots and umbrellas of the treaty-ports [had] not yet appeared,' towns and cities where interaction with foreigners was the norm, quickly became bastions of Western fashion.[31] There was no practical reason for adopting Western dress. Leather boots were uncomfortable for people used to wearing clogs (*geta*) or rush sandals, and woolen pants itched.[32] Emperor Meiji, his officials, the police, and the new conscript army all became decidedly 'Western' in appearance. On 25 August 1871, the Grand Councilors (*Sangi*) and Council of State (*Dajōkan*) circulated a private memorial which stated that in support of the national polity, *kokutai*, the emperor should adopt Western-style military dress. Arguing that accepting these artifacts of Western civilization was more than the 'weak-kneed' imitation of Tang-style court attire worn by generations of Japanese nobility, Meiji leaders claimed that because the Emperor, descendent of Emperor Jimmu and Empress Jinmu, who had conquered Korea, was admired by the masses as commander and chief of the armed forces, that it was appropriate for him, as well as those he commanded, to wear Western-style military uniforms. This memorial, it was claimed, should be considered a matter regarding the 'complete reform of out-of-date customs.' It was not an issue of simple appearance. But exactly how the adoption of Western dress should be differentiated from the previous adoption of Chinese court attire was not explained.[33]

Nonetheless, the official 'uniform' of minister and bureaucrat alike, was the swallowtail coat.[34] Regardless of cost, availability, or comfort, Japan's new army was also quickly outfitted in woolen uniforms and leather boots so as to appear European.[35] The government imported so much leather for boots and saddles that one commentator suggested a change to elephant hide as a cost-cutting measure! Guards at the Imperial Palace resembled the Zouaves at the Tuileries and marched to French light infantry bugle calls, while marines went on parade to the sound of fife and drum in the tradition of their British counterparts.[36] The champions of cultural materiality even saw the Japanese diet as somehow deficient. Lending new meaning to the phrase 'you are what you eat,' they viewed cows as stoic animals with great endurance, and in turn promoted eating beef as a way for the Japanese to increase their stature and stamina.[37]

Japan's new 'civilized' identity was cast in terms of Western technological achievement. Describing the economic and political necessity for adopting some aspect or artifact of the West, Meiji officials and other cultural materialists delineated the benefits in terms of Japan's 'progress,' 'civilization,' or national prestige. Reporting on the government's plans for funding its latest efforts at 'civilization building,' one *Tokyo nichi nichi shinbun* commentator observed how the development of Japan's industry and infrastructure would benefit the nation. By promoting industry, railroads, the telegraph and postal systems, Japan's economy would grow – all the while 'strengthening the fundamentals of a *rich nation* and *strong army*.'[38] Once the people began to work in manufacturing, adopting the machinery and factories of the West, they would become educated [in the 'civilized' ways], support economic development and 'provide the basis for a 'civilized and enlightened country.'[39]

For public works officials like Sano Tsunetami, Japan's adoption of European silk reeling machinery and methods would serve a number of purposes. He claimed that the government's reason for improving Japan's silk reeling methods was to serve the public good. An accompanying increase in raw silk prices and the value of Japan's exports would also follow. This, he stated, was the way things should be – Japan ought to be number one.[40] Sano also revealed the government's other motives in the adoption of Western technological artifacts – elevating Japan's international status. 'The most important change in silk manufacturing these days is the adoption of European methods. The intended result of this plan is, of course, to surpass the other Asian countries. Tomioka filature is evidence of this.'[41]

Sano was not simply referring to the quality of Japan's raw silk when he spoke of Japan's ascending position in Asia and the world. Regarding Tomioka and the adoption of 'superior French methods,' he described the national prestige that the filature had brought Japan. Not only had Japan's level of 'civilization' surpassed the rest of Asia, its machine-reeled silk was evidence that Japan could claim technical superiority over Spain and regions of France and Belgium. 'If one examines Japan's raw silk,' he stated, ' it is superior to that of Cévennes, Spain, and Bruges. As for adopting these countries' methods of manufacture, on the contrary, [we have] surpassed them.'[42] To the Victorian world, technological achievement was 'civilization' and Meiji cultural materialists well understood this 'fact.'

Establishing a country's place in the hierarchy of nations based on technological prowess began, as a popular trend, at the London Crystal Palace Exhibition in 1851. Japan first witnessed such an event under the tutelage of British minister and Japanophile Sir Rutherford Alcock in London in 1862. Leon Roches similarly sponsored the Shogunate's participation in the 1867 Paris exhibition, where Shibusawa Eiichi and Sugiura Yuzuru, among others, observed Western techno-logical superiority first-hand.[43] Would-be cultural materialists became fully indoctrinated in the ideologies of a techno-civilizational hierarchy and the progress ideology of materials at these exhibitions. To ensure that the country made a favor-able impression at the International Exhibition in Vienna six years later, Ōkuma Shigenobu had his finance ministry allocate over ¥600,000 to support the project, which took over two years of planning and preparation.[44] For Japan and the cultural materialists, Ōkuma's investment paid off. In Vienna, the event which provided the bluster for Sano's observations, cultural materialists' beliefs were not only confirmed, they were also advanced.[45] The Meiji government as a whole recognized that the latest technological artifacts displayed at the exhibitions represented modernity. After the Iwakura Mission visited the Vienna exhibition, officials recognized the true value of these events. Not only did they, especially Sano, feel that it was necessary to copy what had been seen at the exhibitions, holding exhibitions in Japan would also be a way to 'modernize.' Scrolls 82 and 83 of Kume Kunitake's record of the Iwakura Mission contain his impressions of the Austrian exhibition. Sano's views were popularized within government circles as well. In an opinion piece, *ikensho*, written for presentation to the Council of State and various ministries, Sano presented a number of reasons why Japan should participate in international exhibitions and hold its own national- and prefectural-level events. Sano's position is significant because he identified the core value of European exhibition thinking: to display goods in an effort to differentiate 'superiority' from 'inferiority.'[46] Not only were cultural materialists able to promote the belief that Western machines and certain materials represented 'progress' and 'civilization' in general, Japan as a country would embrace the exhibitions themselves as evidence of 'genuine enlightenment' and a symbol of a new internationalism.[47] American and British observers in Tokyo agreed. They considered official and popular promotion of industrial and 'scientific' exhibitions as evidence of Japanese superiority over China and the 'other Oriental nations.'[48]

Reacting to the extremes of cultural materiality, officials at the Ministry of Public Works' filature were incensed at the influence that this dominant form of industrial promotion had on local silk reelers. Visitors to the ministry's Akasaka filature were critical of the *wooden* machinery, as well as the *wooden* building in which it was housed. They cared little that the trainees were producing high-quality silk on the simple-yet-effective wooden reeling frames. Although reelers and managers who were trained at the facility efficiently learned many aspects of silk production, filatures throughout Japan continued to fail, and trainees continued to leave Akasaka. Managers at the facility identified the problem as 'the base people who come [to Akasaka] believing that the basic wooden machines are cheap and unsophisticated.' 'Yet in all actuality,' they continued, 'they are not.'[49] Many reelers who trained at this facility returned home only to continue reeling by traditional Japanese methods. Few wanted to invest, or admit to investing in, Akasaka's 'unsophisticated' wooden technologies.[50] The program which over-popularized Tomioka's unattainable, steam-powered, cast iron technologies helped foster an attitude which contributed to the demise of the Akasaka filature and a relatively unbalanced adoption of Western reeling methods and techniques. In essence, the cultural materialists were self-defeating on at least one front. By urging only the adoption of 'superior' Western technologies, they inhibited the spread of the viable alternatives, thus placing artificial restrictions on the development of the silk reeling industry – contrary to one purported goal of early Meiji industrialization.[51] Appendix II gives details of distribution and productivity for all filatures in 1895.

Iron's role in *bunmei kaika* and 'civilization building' was particularly significant. Imported industrial artifacts, sturdy bridges, 'earthquake resistant' lighthouses, and railroads were all constructed from iron.[52] During the first period of Meiji modernization, however, iron should be considered as having more of a supporting, rather than a leading, role. Iron was the material from which to build 'civilization'; it was the medium through which a progress ideology of materials was transmitted. As such, the government appropriated and opened iron mines and invested in an iron industry as part of its efforts at 'civilization building.' Officials argued that the amount the government spent on an iron industry directly related to its 'chosen level of civilization.'[53] However, nearly all Meiji era iron needs had been satisfied through the use of imported iron; and this was the issue which forced open the government's purse. Japan's efforts at 'civilization building,' as well as those aimed at providing the basis of a *rich nation* and *strong army*, it was reasoned, could not rely on imported iron.[54] As a result, by 1880 the government had spent well over ¥2,000,000 promoting the iron industry in its effort to make Japanese 'civilization' reliant on indigenous sources of this raw material.[55]

Iron import substitution and the Meiji government's general interest in iron only came about with the founding of the Ministry of Public Works in 1870, and the subsequent increased demand for the material following new policies which called for building railroads, a telegraph system, and iron bridges over Japan's numerous streams and rivers. Iron was also necessary to supply the ministry's new factories, such as the Sapporo Machine Works (1875) and Akabane Engineering Works (1874). Based on American and British designs respectively, these facilities were established to produce steam engines, machinery, and other icons of 'modernity.' These artifacts, the hallmarks of 'civilization,' also relied on imported iron.[56] A modern navy similarly required large amounts of iron (and steel), and this was also part of the impetus for founding a national iron industry. Yet Kamaishi was sold to a private buyer despite the argument that Japan should not rely on foreign iron.

Even the government's great investment in the industry was not enough to prevent the sale.[57] But iron was iron, imported or otherwise; and as long as Japan was able to lay track, make machines, and build in iron, the cultural materialists' progress ideology of materials was satisfied.

Curiously, neither Kamaishi ironworks, nor any other ironworks in Japan for that matter, was ever described in terms of *bunmei kaika* or 'modernity.' While Kamaishi's British engineers frequently mentioned the fact that the government's ironworks was outfitted with the most modern equipment, the government never referred to the facility in the same way. The same cannot be said, however, for iron *artifacts*. Unlike their wooden counterparts, cast-iron machines were frequently described as 'sophisticated' or 'modern.'[58] Some officials, like Sano Tsunetami, even went to great lengths to indicate perceived differences between traditional or wooden and modern iron machines through his selection of the Chinese characters (*kanji*) used in describing different industries and equipment. The Japanese word for machine, *kikai*, can be written two ways while maintaining the same pronunciation by simply changing the first character. When discussing traditional industry, Sano used the character *utsuwa*, which had the accepted meaning of an appliance related to wood or cutting wood. For iron machines, mechanized industry based on Western principles, and mining he used the character *hata*, which meant machine or device.[59] Although not common to all Meiji texts dealing with industrial matters, Sano's explicit linguistic differentiation between traditional and imported 'modern' machines and industries, combined with his comments about industrial promotion, its ties to 'civilization,' and his recommendations for Japan's industrial future, are revealing of the progress ideology of materials that was pervasive among so many Meiji officials and intellectuals during the first period of Meiji *techno-identity* formation.

The second phase: selective restructuring

Japan's second phase of *techno-identity* formation, from roughly 1878 until the early- to mid-1890s, can be described in general as a turn from the unbridled, yet purposeful, acceptance of *materia Hesperia* to further foreign policy goals, to a time of *selective techno-political restructuring*. During this period, many of the government's actions with regard to the acceptance or rejection of Western cultural icons were representative of general shifts in the way Meiji leaders and intellectuals viewed Japan and its relationship with the world. To say that Japan was turning an imperialistic eye toward Korea is largely a given. And although we still have not reached consensus on why Japan sought to expand onto the mainland at this early date, the theory that Meiji officials wanted to spread 'civilization and enlightenment' to the 'hermit kingdom' is largely considered part of the story.[60] Through diplomatic action and the outright threat of war, Meiji officials abandoned Japan's former identity of samurai warrior doing battle with Korea in the name of national honor, as was purportedly the case earlier in the decade, and began the process of redefining Japan as a Western-style diplomatic and military power for the sake of national security.[61] The problem, however, was that Japan did not yet have the bite to match its bark. While there were still some officials and intellectuals who continued to support 'civilization-building' efforts within Japan as part of foreign policy, other Meiji leaders and intellectuals began to redefine Japan in terms of a new set of technological artifacts – ones with greater military importance.

Efforts to construct Japan's new identity at all levels collided with – or were part and parcel to – conflicts within the government. The sale of government enterprises, for example, typically considered a matter of economic policy, was equally a product of inter- and intra-ministerial conflict, and a growing conservatism in the Meiji bureaucracy in general. Disposal of crucial 'civilization building' enterprises was politically situated within the 1881 political crisis, the so-called Matsukata Deflation, and *hanbatsu* (domain alliance) rivalries. At the same time, the Meiji government was growing politically more conservative. Japan was keenly aware of the advances of Western imperialism, and was beginning to venture more energetically into empire-building. These changes can be seen as reactions to increasingly popular liberal trends, and the government's recognition of its own domestic position and Japan's rising international status. Just as the importation of Western cultural icons had been the prerogative of a relatively small group of cultural materialists, their rejection was the product of different political and intellectual commitments expressed by a core group of government officials and intellectuals. As such, the sale of government enterprises should not be considered strictly an economic event. Rather, it was inextricably tied to Japan's evolving (domestic and international) identity as expressed in technological artifacts.[62]

Any discussion of Japanese industrial policy or technology in the early- through mid-1880s invariably focuses on the government's disposal of its so-called model factories and enterprises as part of Minister of Finance Matsukata Masayoshi's deflationary policy. The sale of government industry, *kōgyō haraisage*, was, however, a plan that predated Matsukata's rise to this all-powerful office. In fact, plans to sell select facilities had been under consideration since at least 1876,[63] and many argue that this was the government's intention from the very beginning. What never seems to have been an issue, however, was the type of industry which the government intended to sell. Each time some official suggested disposing of the government's factories, (silk, cotton, and wool) textiles, glass or metal machine manufacturing, and even brick and cement works were targeted. No one identified the government's (iron) mines or refineries as white elephants, at least not until Matsukata became finance minister on the heels of the political crisis which removed Ōkuma Shigenobu from the government in October 1881.[64]

As Minister of Finance, Ōkuma had a number of potential remedies for the government's mounting financial woes.[65] After narrowly escaping political suicide when the emperor rejected his plan to borrow 50 million yen from England to finance the conversion of Japan's increasingly worthless paper currency, he once again turned to expenditure reduction and the sale of government industries.[66] In a proposal written in May 1880, Ōkuma outlined his reasons for urging the sale of the public works and home affairs ministries' factories.[67] Mostly he argued that the facilities had served their purpose: industrial promotion in areas that were possibly too risky or expensive for timely action by Japan's seemingly fiscally conservative entrepreneurs. The time had come, he argued, for the government to turn over its enterprises to the people. Most all of the facilities were operating in good order, and it was time for the people to have the opportunity to reap the profits from these enterprises before they became 'government monopolies.'[68] Looking at the proposed sale from the government's perspective, Ōkuma argued that no good would come of things if the country's financial resources remained tied up in the factories. The need for funds to repay the national debt was becoming increasingly urgent, and at the very least, he argued 'we would not be remiss in curtailing expenditures.'[69]

Ōkuma targeted for disposal some 14 facilities operated by the Ministry of Public Works and Ministry of Home Affairs, including the Senjū Woolens Mill, two cotton spinning mills, the Shinmachi Waste Silk Reeling facility, Tomioka Silk Filature, a sugar refinery, the Akabane Factory, and the Fukagawa Brickyard.[70] Under no uncertain circumstances, however, was the government to sell its mines, refineries, or ironworks. These industries, Ōkuma reasoned, required high-levels of training and financial support and 'could not be left to the private sector.' The transformation of raw material into artifact at the government's arsenals was similarly a process that would be 'done for the people.'[71] In addition to requiring even higher levels of training, a certain degree of secrecy, he reasoned, was essential at the government's arsenals – as well as at its publishing houses.[72]

Inoue Kaoru, then Minister of Foreign Affairs, similarly issued a proposal to the Council of State outlining his ideas for disposal of the government's enterprises. He observed that Japan had made important 'progress' through the importation of the technical achievements and artifacts of the West. Creation of the basis of a '*rich nation*' – railroads, steam ships, telegraph, and lighthouses – was a praiseworthy achievement. Japan had also made 'progress' in silk reeling, cotton spinning, wool production, and glass, paper, and iron refining/manufacturing.[73] Inoue argued that the government should gradually reduce its expenditures by selling these facilities to the private sector, but that any attempt rapidly to dispose of the government's enterprises would be counterproductive.[74] Significantly, Inoue, like Ōkuma, never mentioned mining or disposal of the government's ironworks.

On 5 November 1880, less than two months after Inoue presented his memorandum, the Council of State published the 'Regulations Governing the Sale of Factories' (*Kōgyō haraisage gaisoku*), which outlined the terms for disposing of state-owned enterprises.[75] This document was accompanied by another which ordered the Ministries of Finance, Home Affairs, Army, Navy, Education, Public Works, plus the Hokkaido Development Commission to reduce expenditures. While fairly specific regarding terms of sale and degree of future government involvement with the factories, there is no indication of what types of enterprises the government intended to sell. In the main, the government's directive ordered that no new projects for industrial promotion be undertaken, and that previously established enterprises should proceed to operate in a more fiscally cautious manner.[76]

If Ōkuma's and Inoue's recommendations are an indication, Kamaishi Ironworks and its associated mines were not included in the disposal plans. That no one recommended selling the government's mines, smelting facilities, and refineries should come as no surprise. At the very least in theory, mines and their associated refineries should be profitable – especially in the case of precious metals and coal. The government's ironworks had the potential for profit, but profit was not the reason so often cited as to why Japan needed to be iron self-sufficient; profit does not even appear to have been a consideration.

Iron was a 'strategic' industry for the Meiji government. As early as 1872, iron was identified by Meiji officials as the material from which to build 'civilization.'[77] Throughout the first decade after the Restoration, government officials continued to recognize that iron – in the form of machines, railroads, bridges, and the like – exemplified Japan's 'level of civilization.' And 'progress toward civilization,' argued officials like Yamao Yōzō and Itō Hirobumi, must not be reliant on imported iron.[78] Japan's *techno-diplomacy* and efforts to renegotiate the unequal treaties could not rely on the economic whims of Great Britain or the United States.

While it does not appear that the government intended to sell Kamaishi at this early date, there are indications that some officials were hoping at least to re-appropriate or cut some of Kamaishi's funding. A jointly signed petition issued to the Council of State in late February 1880 (i.e., four months before Ōkuma's proposal and ten months before Kamaishi's furnaces were tested and blown-in) by the Ministries of Army, Navy, and Public Works briefly outlined plans which argued for the continued funding of a 'large ironworks' on a cooperative, inter-ministerial basis. The tenor of this petition was surprisingly similar to the original memorandum that Yamao and Itō had written almost a decade before. The future of Japan depended on iron because so much of the 'modern' world relied on this raw material; Japan must abandon the 'evil practice' of dependency on foreign imported iron; and the iron industry was described as beyond the capability of private enterprise.[79]

There was, however, one significant difference between this petition and those which it preceded: the technological artifacts identified as crucial to the nation and Japan's identity. Reflecting a changing world view and a growing sense of national assertiveness, the authors of this petition did not identify iron machines for economic development or railroads as icons of Japan's *techno-diplomacy*. Rather, their focus was more indicative of the nation's growing international involvement and Meiji officials' understanding of a Victorian era world during the period of high imperialism. 'Iron is required for everything from *warships* and *their engines*, to both *large armaments* and *small arms*, to all kinds of machinery. In general [Japan] cannot do without iron.'[80] By redefining the importance of iron in terms of military preparedness and national security, Meiji officials had unknowingly initiated the decline of *bunmei kaika* materiality.

The end of *bunmei kaika*

Rejection of the 'civilization and enlightenment' movement is commonly placed at the beginning of the1890s. In his study of Nishimura Shigeki, an early proponent of 'civilization and enlightenment' turned Confucian moralist and educator, Donald Shively argued, for example, that "Civilization and Enlightenment' died with the promulgation of the Constitution of 1889 . . . and the Imperial Rescript on Education of the next year.'[81] Kenneth Pyle assigns the end of *bunmei kaika* to the outbreak of the first Sino-Japanese War in 1894.[82] While it may have taken until the 1890s for *bunmei kaika* to die, the ideology had, in fact, been terminally ill for at least a decade.[83] By 1880, the focus of government-sponsored industrialization efforts was no longer 'civilization building': industrial promotion was now framed in terms of 'military preparedness' and 'national security.' Gone from official statements on industry was the rhetoric of 'civilization and enlightenment.' Breaking the long silence, official reports of iron production at Kamaishi only slowly began to appear in the popular press. These articles, unlike frequent coverage of Japan's 'progress toward civilization' at Tomioka Silk Filature, had nothing to do with 'civilization building' – they presented a new perspective on Japan's technological development and drive for iron self-sufficiency.[84] One article published in December 1880, for example, argued that Japan's prowess in iron manufacturing made it the technological equal of the West. Japan's identity was being recast in terms of high-quality iron. The author stated that upon testing, iron refined at Kamaishi Ironworks was of equal quality to that which had been imported from Great Britain. The article goes on to state that increased domestic production and the end of reliance on the West for iron was 'a cause for celebration.'[85]

As held true for Sano's observations at the Vienna Exhibition in 1873, technological achievement was still the yardstick by which the Victorian world measured 'progress' and 'civilization.' For some, Japan was, or was becoming equal to Britain. In redefining Japan's technological identity, however, a new generation of conservative commentator began to promote the Japanese nation as separate and distinct from the mono-cultural entity that was part and parcel to the rhetoric of 'civilization and enlightenment' theory.[86] While still in its nascent stages, and far from being universally accepted, by 1880, Meiji officials and intellectuals began selectively to restructure Japan's identity in terms of a new set of technological artifacts, ones that would help redefine Japan as a naval power.

In an ironic twist, Japan's new technological identity was more fully cast in iron *following* the sale of the government's ironworks. As a result of Kamaishi's sale, numerous government officials called once again for the establishment of a national ironworks. While serving in France under prefectural orders in 1887, Doctor of Engineering Obana Fuyukichi argued in his 'Discourse on Establishing an Iron Industry' that it was necessary for Japan to develop its iron-producing capacity.[87] Obana stated that his report would 'illustrate the degree of civilization in various countries through an examination of comparative norms of military preparedness, transportation, and manufacturing; all points related to iron production and the standard of progress in the iron industries [of various nations].'[88] Basing his study on information collected from British engineering journals, Obana argued that 'a large ironworks would provide the conveniences of civilization (*bunmei no riki*) in times of peace.' More importantly, however, the same ironworks 'would provide the necessary tools for defending the country in time of war.'[89] To support his claims, Obana compiled production statistics for pig iron and steel production in Great Britain, the United States, France, Germany, Russia, Austria, Belgium, Switzerland, and Italy. He then correlated this data with each country's naval tonnage, miles of railroad track, and general shipping tonnage. To demonstrate where Japan ranked in the world hierarchy of iron, he added British India, Turkey, Brazil, Mexico, Canada, Spain, and Portugal to the list.[90] Although the data do support his conclusions, Obana ranked Japan with or above Italy. In fact, Obana reiterated this point at each stage of his comparison.[91] For this engineer, Japan had become a Western-style power. What is clear from his 'Discourse' is the importance of iron for armaments and warships. Throughout the report, he reminds his readers that for a nation to be militarily strong it must have a sizable iron industry. His conclusion is that Japan, already a 'civilized' nation ranking among the countries of Europe – and well above 'primitive' countries like the Ottoman Empire and Mexico – was well on its way to becoming a world power. Japan needed more iron, however, if it was to continue making 'progress,' and this necessitated the construction of a large national ironworks.

Entering the Ministry of Finance in 1884, Soeda Juichi also saw iron as essential to Japan's national security. As a maritime nation, Soeda similarly cast Japan's new identity in terms of being a naval power. In a paper entitled 'Building an Ironworks is Urgent Business,' Soeda argued that Japan must expand its iron industry and become self-sufficient in order to increase the size of its navy and armaments industry.[92] Interestingly, Soeda both described Japan in terms of Britain and also in opposition to that country. He stated that as a maritime nation Japan must, like Britain, be self-sufficient in iron production – even in times of war. Conversely, he argued that Japan should not be beholden to the other island nation for the raw materials with which to build armaments, a navy, and related machinery.[93]

Obana's and Soeda's writings on the necessity of establishing a national iron industry reflect a growing sense of urgency in the midst of the political and economic debates which surrounded continued efforts to fund Kamaishi and, in Soeda's case, proposals for establishing a new government ironworks at Yawata in 1892. Japan's evolving *techno-identity* and the changing cultural value of iron was being reformulated in terms of 'military preparedness' and 'national security' while some Meiji leaders and intellectuals were articulating a more aggressive policy toward Korea based on the notion that an 'enlightened' and independent Korea was of strategic importance to Japan in fending off Western imperialist advances.

China's efforts to reassert authority in the region were also interpreted as threats to Japan's national security. At the same time that the Ministries of Army, Navy, and Public Works were reminding the Council of State and the Ministry of Finance of the importance of an iron industry for Japan's national security, the Chinese Governor-General of Zhili, Li Hongzhang, was urging Korea to throw open its doors to the Western powers in a move designed to limit Japanese influence on the peninsula. By opening Korea to (Western) trade and diplomatic activity, Li reasoned, Korea would survive the onslaught of foreign aggression while the Western powers and Japan vied for position.[94] Real or not, Li's efforts to build a modern Chinese navy were similarly seen as a source of future aggression by Japan. By 1882, Li had managed to amass a small fleet of steam-powered warships – half imported from Britain, the other two dozen or so of local manufacture – some of which, according to contemporary observations, were quite well armed and modern.[95] Efforts to rein Korea back into the Chinese world order were not simply perceived as slights to

Japan's national prestige as had been the argument a decade before.[96] 'Civilizing' Korea, as apart from China, was perceived as essential to Japanese national security and its ability to maintain peace in the region. The Meiji government's capacity to project Japan's image in the international arena would rely on a large navy and thus, as had been argued, a steady supply of iron.

Where once Europeans (and Americans) were the primary threat, China had, in the 1880s, become Japan's 'chief hypothetical enemy.'[97] Recognizing China's potential to field a massive army, military leaders like Yamagata Aritomo tirelessly petitioned the government to increase military spending in an effort to loosen bureaucratic purse strings. Civilian leaders such as Iwakura and Itō only recognized the 'validity' of Yamagata's argument after the 1882 Imo Mutiny, which followed botched Japanese attempts to 'export enlightenment' to Korea. Japan would be able to neither launch offensive actions in Korea nor defend itself from Chinese attack without increasing the size of both its army and navy.[98]

A growing conservatism within the Meiji government and society contributed to the propagation of Japan's new attitude toward China, Korea, and the West, its rejection of 'civilization and enlightenment,' and concurrently changing national identity.[99] Not all government officials, and certainly not all leaders of society, were becoming conservative. There were still a large number of *bunmei kaika* cultural materialists who gauged foreign policy in terms of Japan's Western appearance.[100] The mere presence of such liberalizing forces as popular rights movements or calls for the county's outright Westernization in some cases, helped to create the conservative, anti-'civilization' crusade which worked toward resuscitating Confucian moralism and undermining the People's Rights Movement.

Japanese conservatism did not grow in isolation; it can be seen as part of an international conservative trend that some scholars have called 'an era of Conservative supremacy.'[101] Following Carlton J. H. Hayes' description of Europe during the 1840s through 1870s, where the 'era of Liberal ascendancy had witnessed a marked decline of European imperialism,'[102] Japan's era of ascending liberalism of the 1860s and 1870s, *bunmei kaika*, was similarly marked by a policy which can be characterized as national consolidation. Japan's punitive expedition against Formosa, its abandonment of Sakhalin and claim to the Kurile Islands, and the 1876 political crisis over Korea, can all be seen as expressions of national boundary definition and of establishing protective zones. Entering the 1880s, however, Japan, like Europe (and the United States) of the same decade, launched a course of nationalism *cum* imperialism that would continue through the end of the century. But Japan was not Europe, and where Western actions can been interpreted as economic imperialism – the search for raw materials and new markets for Western mass production, the same cannot necessarily be said of Japan. Japan was fending off the West while simultaneously trying to become more like its rivals. Through its efforts to renegotiate the unequal treaties and through diplomatic and educational missions sent to Europe at the time, many Japanese began to question the utility of 'becoming naturalized Westerners.'[103]

For at least part of the Meiji bureaucracy iron was the raw material from which to continue building a '*rich nation*' and '*strong army*,' or perhaps more appropriately

Figure 5.1 Katakura & Co. Filatures raw silk label 'Ammunition Brand.' Even silk reelers capitalized on the Meiji identity defined in terms of military preparedness and national security (*Source*: Author's collection)

a '*strong army*' and '*rich nation*.'[104] National security, military preparedness, and greater economic independence all relied on iron. One could even argue that iron self-sufficiency and the accompanying necessary increase in arms production, a factor given in almost all pro-iron industry development reports, also meant iron was the means by which to secure Japanese national independence.[105] If iron was this important for Japan, and Kamaishi Ironworks was not on the list for disposal by

Figure 5.2 Katakura & Co. Ltd. 'Propeller Brand Raw Silk', c. 1905. Continuing the trend
of using the symbolism of military preparedness, this label makes specific
reference to naval imagery (*Source*: Author's collection)

virtue of its status as a strategic industry, the question of why officials sold the government's mines and ironworks requires closer examination.

In strictly economic terms, Kamaishi was expensive. By 1881, when Matsukata came to power, the Meiji government had invested nearly two million yen in the ironworks and had little to show for its efforts.[106] But Kamaishi was not sold purely for economic reasons. As argued in Chapter 4, iron did not have the same cultural implications for 'civilization building' as did other Western material icons. Iron was not even iconic per se, it was the material from which icons were made. Although it was an integral part of *bunmei kaika* materiality and was rapidly becoming part of Japan's new identity cast in terms of military preparedness, Finance Minister Matsukata's personal view of the large-scale iron industry appears to have been one factor which sealed Kamaishi's fate as a government enterprise.

For at least a decade prior to Kamaishi's sale, Matsukata had been an ardent supporter of agricultural promotion and the development of Japan's agro-industries. He believed that the government should use its financial reserves to improve the land, help farmers develop better cultivation techniques, and fund animal husbandry projects. He argued conversely, that the Meiji government should neither emulate foreigners by importing their machines nor promote or fund 'Western-style industry,' by which he meant factories or mines.[107] In support of this position, Matsukata authored numerous opinion/policy papers dedicated to articulating his views on improving Japan's agriculture. These works include 'Recommendations for a Compilation of Agricultural Books' (*Nōsho hensan no gi*), in which Matsukata compiled Japanese and Western agricultural treatises to use as a guide for promoting 'scientific agricultural methods,' and 'An Outline of Policies to Encourage Agriculture' (*Kannō yōshi*) which criticized the present state of Japanese agriculture.[108] He targeted nearly a dozen areas in which Japanese agriculture and agro-industries were lacking including the negligible progress made in livestock rearing and dairy farming, and the slow pace at which farmers were willing to try 'modern' methods of agricultural production. He also laid out a number of 'solutions' within a general framework of promoting indigenous goods as well as those in which he saw the promise of profit. Matsukata was also an ardent supporter of agricultural exhibitions designed to increase popular awareness and reward those who adopted 'scientific methods.' As if foretelling the Council of State's 1880 memorial, Matsukata recognized both the importance of government efforts to promote agro-industry, and of turning successfully established enterprises over to the private sector.[109]

Matsukata was a self-described proponent of laissez-faire liberalism.[110] He regularly recommended against official involvement in the promotion of heavy industry or industries which relied on imported Western machinery, citing earlier government experiences as evidence in support of his position. In an 1890 report which discussed Japan's earlier currency dilemma, for example, he detailed the problems of the Meiji government's investment in industry, and how earlier efforts to promote the Westernization of Japanese industry had only produced debt.[111] But his economic ideology was more complicated than simple laissez-faire. It would be more accurate to describe the finance minister as believing in only one 'invisible hand,' that guided Western-style industry and economics; for agro-industry

Matsukata's hand was clearly visible. From 1880 through the end of the century, government subsidies for mining and manufacturing were quickly reduced, while those for agriculture, forestry, and fisheries gradually increased.[112] The bulk of government funding was directed toward public works projects like flood control and reforestation, programs relating directly to the agricultural sector.[113] At the same time Matsukata was disposing of Kamaishi Ironworks, the building block of national security, he was spending over three million yen on agricultural promotion – over ¥980,000 in total on unsuccessful livestock-related projects alone.[114]

Much as Brunat had recommended at Tomioka, Matsukata believed that Japan's industrial base, by which he meant agriculture-related industries such as silk reeling or tea, should rely on indigenous technologies that would gradually be improved through the incorporation of useful foreign techniques.[115] He was very much against rapidly importing Western technological systems, the symbols of 'modern' industry, as the basis for Japanese economic development. Unlike the cultural materialists who planned Tomioka, Matsukata can be seen as a *techno-conservative* – one who based technological development on the gradual incorporation of proven artifacts on an individual basis. In one sense, the finance minister was pragmatic: Japan's paper currency was largely worthless and its specie reserves were growing dangerously low: Matsukata needed to reduce expenditures and shore up Japan's economic base. He was not opposed to Western technological artifacts; indeed, he recognized the importance of some, such as the railroad for Japan. He just did not want to pay for them.[116] But Matsukata's extreme devotion to the agricultural sector and his willingness to abandon heavy industry made him either incredibly principled or very short-sighted. At the same time he was devising plans for the privatization of Japan's railroads and trying to find a buyer for Kamaishi Ironworks, he was importing 'Shropshire and Southdown sheep and [placing] them in special breeding pastures designed for their needs.' The sheep did not fare well, but this did not stop the Ministries of Finance and of Agriculture and Commerce from spending 'hundreds of thousands of yen' on the project.[117]

Published in 1884, and often considered the grand plan behind Japan's second surge of industrialization which began prior to the first Sino-Japanese War of 1894, *Kōgyō iken* ('Views on the Promotion of Industry') illustrates some of the inconsistencies behind Matsukata's views on industrial promotion.[118] Although largely the efforts of Maeda Masana, Matsukata's protégée and a developmental pragmatist in his own right, the official version of *Kōgyō iken* also illustrates the finance minister's position on a number of issues.[119] In the second volume, the authors outlined the importance of a strong military which must be based on domestic production and capital. Japanese national security should not depend on foreign loans or imported weaponry.[120] While much of the discussion identified the martial spirit and bravery of Japan's 'great generals and soldiers' as the backbone of the nation's military, Maeda compared European and American military expenditures in an effort to draw a correlation between a country's ability to succeed in war and its economic base. He concluded that the 'power to win wars is in one's soldiers and one's armaments' and 'Japan needs to be militarily prepared.'[121] In order to be militarily prepared, Japan needed iron and steel.

Kōgyō iken is peppered with comments which reiterate that its authors were thinking about strengthening Japan's military and industrial might.[122] In fact, part of the argument for expenditure reduction and the sale of government enterprises was that this would make it possible to increase funding to the military. But how Japan would accomplish the task of building a modern navy and producing its armaments without being dependent on foreign sources of iron and steel never seems to have been considered.

When Kamaishi first started producing iron in 1880, the Meiji government was importing more that 40,000 tons of iron and steel annually. Imports increased the following year and then dropped only slightly, to approximately 35,000 tons (31,752 tonnes), during the 'deflation years.' In 1885, the year in which the so-called Matsukata Deflation 'ended,' the demand for iron and steel jumped back up over the 40,000-ton (36,288 tonnes) mark; the following year annual imports topped 60,000 tons (54,432 tonnes). For the rest of the Meiji era, Japan's dependency on

Table 5.1 Iron and steel imports, 1874–1900 (weight in long tons, value in pounds sterling)

Iron and Steel Imports

Year	Amount	Value
1874	11,368.20	86,480.90
1875	15,249.30	94,554.20
1876	10,999.10	72,141.10
1877	17,097.40	97,341.50
1878	24,512.00	148,679.80
1879	24,810.50	127,442.30
1880	37,109.80	186,098.90
1881	38,467.60	153,282.90
1882	32,302.20	134,951.60
1883	30,236.00	120,052.00
1884	32,117.50	114,084.00
1885	39,151.00	141,152.80
1886	52,085.50	164,567.40
1887	65,477.00	222,065.60
1888	107,166.00	366,067.70
1889	73,107.00	314,383.30
1890	78,352.00	412,239.10
1891	71,231.00	311,475.70
1892	48,802.90	234,531.50
1893	85,894.00	374,074.10
1894	124,960.00	690,158.60
1895	137,514.00	819,339.70
1896	231,124.00	1,158,145.50
1897	249,511.00	1,346,198.40
1898	297,868.00	1,508,811.40
1899	147,470.00	1,237,934.70
1900	272,368.00	2,720,046.40

Source: Based on *Bureau of Mines*, 1909: 57

foreign iron and steel remained at nearly triple the 1880 levels.[123] On average, from 1880 to 1885 the government spent £141, 604 per year importing iron and steel. From 1886 through 1890, £298,467 per year, and £597,954 annually from 1891 to 1895.[124]

In an effort to parry some of the criticism received for their inability to keep Kamaishi in-blast, public works officials explained away their problems in terms of poor-quality iron ore. The logic of their excuses, not to mention the fallacy, did not even survive public commentary. One month after Kamaishi's furnaces were restarted in the spring of 1882, an editorial in the *Jiji shinbun* lambasted the Ministry of Public Works for claiming that they had been unable to locate high-quality ore in the Kamaishi region. With all that had been invested in the project, how was it possible that Kamaishi's pig iron was so inferior as to render the 'newly imported machinery useless?'[125] Kamaishi's critics were correct. The ironworks' problems, at least until this point, had little to do with the quality of raw materials.[126]

When Itō Yajirō, a Ministry of Public Works inspector, visited Kamaishi later that year, Matsukata had already identified the ironworks for disposal. Problems at the facility were described as 'being beyond words.'[127] Itō condemned the ironworks based on his assessment that there was not enough iron ore at the site. He claimed that there were only about 130,000 tons (117,936 tonnes) and contended that most of it was located in rough, steep terrain that would be uneconomical to mine.[128] Kamaishi's history has demonstrated that Itō was incorrect in his appraisal; but even at the time, people questioned his assessment. One beneficiary of Matsukata's sell-off plans, Asano Sōichirō, the new owner of the Fukagawa Cement Factory, questioned not only Itō's report, but the haste with which the government was trying to dispose of Kamaishi.[129] Asano, Abe Kiyoshi, and a section chief from the Engineering Office went to Kamaishi to inspect some of the machinery that was offered for sale.[130] Having been told that the mines were abandoned because there was no iron, the men thought that they might find something of use. Asano, for example, was interested in finding equipment for drying clay at his factory.[131] After examining the furnaces and blowing engines, the three men followed the rail line into one of the mines. Asano noted how he found it hard to breath in the dark, dank space but decided to return the following day for further inspection. The next morning, Asano and Abe were back in the mine when Asano noticed what appeared to be a 'rusty iron ore-like material flowing from cracks in the rock face.' Remembering that he had been told that there was no iron at Kamaishi, Asano was ecstatic at his find; Abe, however, found the entire episode and Asano's enthusiasm amusing.[132]

Upset at what he considered irresponsible gossip about government expenses and Matsukata's deflationary plan, Asano decided that he had to settle the Kamaishi matter immediately. He visited Shibusawa Eiichi just after New Year's Day 1884, and shocked the former finance ministry official and president of the Dai-Ichi Bank with his insistence that Shibusawa purchase Kamaishi mine. Asano repeatedly stated that there was iron at Kamaishi and that one only need invest ¥100,000 in the government's ironworks to revive the mine and be successful. He even had ideas for buying back some of the components which had already been sold.[133] Shibusawa was intrigued, although declined the initial proposal. He approached Furukawa Ichibei, who had purchased the Ashio Copper Mine from the government in 1877, for advice. Furukawa too, was amused at Asano's persistence and the suggestion that Shibusawa buy the 'worthless' mine. But Asano would not surrender, and continued to argue that there was iron at Kamaishi, which he had 'held in his hand.'[134]

Still unsure of the best course of action, Shibusawa, Furukawa, and Asano consulted Inoue Kaoru, then Minister of Home Affairs. Shibusawa wondered if he should look into the matter because Asano was adamant that the government should not sell Kamaishi.[135] Inoue, however, would not consider giving Kamaishi a second look. He stated that the government had invested large sums of money 'to keep Kamaishi alive' but that the section chief said there was no iron ore. Shibusawa wanted to verify the facts for himself, and thus contacted the section chief, Itō Miyoji, who told him in no uncertain terms that reviving Kamaishi was out of the question. 'Whether ¥100,000 or ¥10,000, it would be like throwing [your money] into a muddy ditch. Please stop thinking about this. The government has already spent [i.e., wasted] ¥2,500,000 like this. Isn't abandoning it the best way?'[136]

Asano had nothing to gain by questioning the sale of Kamaishi since he was not interested in buying it. While he was friends with Shibusawa, however, he seems to have been more concerned that the government was abandoning the ironworks based on erroneous assumptions. On this point, Asano was correct. There was iron at Kamaishi, enough for the ironworks to be economically viable and a commercial success by 1890. On the eve of the first Sino-Japanese War, in fact, over half of Japan's pig iron was mined and smelted at Kamaishi ironworks.[137] Where Asano seems to have been wrong was in believing that the government, that is the Minister of Finance and his supporting Sat-Chō clique, would split ranks and question the disposition of Kamaishi ironworks in the first place.[138]

Prior to Matsukata's ascension to the headship of the finance ministry, one of the few voices of opposition to the Satsuma-Chōshū (Sat-Chō) alliance was Ōkuma Shigenobu. Contributing to the hardening of the Sat-Chō oligarchy against this minister from Hizen was his disclosure and public criticism of what appeared to be the scandalous sale of government properties on Hokkaidō. At the end of the ten-year allocation term for the Hokkaidō Colonization Bureau, which also coincided with the Council of State's 1880 directive to dispose of the government's enterprises, Godai Tomoatsu, former Minister of Civil Affairs, and Kuroda Kiyotaka, Bureau Chief of the Hokkaidō Colonization Bureau, both Satsuma men, entered into an agreement whereby Godai would purchase the government's Hokkaidō properties for ¥300,000, on an interest-free, installment loan to be repaid over 30 years.[139] Although there was disagreement within the government, the deal, as recommended by Kuroda, was approved by the Council of State in July 1881. As soon as news of the proposed sale became public, it was widely condemned by both progressives and conservatives alike.[140] Popular criticism of the apparent scandal raged throughout the summer. By autumn, battle lines within the bureaucracy had hardened, isolating Ōkuma and his political allies. Within two months, in October 1881, Ōkuma would be forced out of the government, but not before seeing the defeat of the Hokkaidō land deal.[141]

With the memory of the Hokkaidō scandal still fresh, it appears that a more conservative, yet cautious, Meiji bureaucracy felt it prudent simply to let Kamaishi fade back into the relative oblivion from which it had briefly emerged. This was easy after all, since the circumstances of its sale were quite unlike those of other government properties.[142] As far as Kamaishi's detractors were concerned, the ironworks showed no potential for profit; and this would also help ensure that no one could cry foul-play. In fact it appeared that no one, other than perhaps Asano, had any interest in reviving the government's ironworks; and officials seemingly wanted to keep it that way.

Kamaishi's sale had all the potential for being another scandal for the government. Tanaka Chōbei, a government supply merchant, was approached in August 1883 by Meiji officials with a 'request' that he purchase Kamaishi. According to what little evidence remains, Tanaka was told nothing about the problems of procuring raw materials at the former government ironworks.[143] The official explanation was that there had been little interest in the facility because of its remote location and a general unfamiliarity with the area.[144] With Matsukata's support and the recommendation of Yokoyama Kyūtarō as an assistant, Tanaka agreed to buy the facility – in its entirety – for a fraction of its original cost.[145] And this transpired at the exact same time that Asano was attempting to convince Shibusawa and Inoue that they should re-examine the proposed sale of Kamaishi.[146]

Just how much of a bargain Tanaka got when he purchased Kamaishi has never been called into question because all of the government's enterprises were purportedly sold at a loss.[147] Compared to the other enterprises, however, Kamaishi Ironworks was literally given away. Kamaishi's sale recouped approximately one half of a percent of the government's investment. Nakakosaka iron mine, a losing proposition from the start, was sold for 33.4 percent of its original value; Asano's Fukugawa Brick Factory, 89.9 percent. On average, the government sold its enterprises at 40 percent of their investment values.[148] In fact, only Kamaishi Ironworks was disposed of for less than 15 percent of the government's original investment. It is no wonder that the government tried to ensure that the sale went all but unnoticed by the public.

That the Minister of Finance, arguably the most powerful Meiji official, gave Tanaka, an Edo hardware wholesaler turned army supply merchant, his personal recommendation is remarkable. Tanaka, however, had longstanding ties to the Shimazu family and, as a Satsuma man himself, had associated with Matsukata and other officials for some time.[149] Just after the Meiji Restoration, in fact, Matsukata was one of the men who approached Tanaka with the recommendation that he make the move from hardware to government supplier.[150] It appears that all the ingredients for another 'land scandal' were in place: Satsuma officials were selling government properties to Satsuma businessmen at rock-bottom prices. The Sat-Chō bureaucracy refused to re-investigate claims that there was iron at Kamaishi, and there was also a letter from Tanaka to Matsukata which confirmed Asano's suspicions. The letter, an official 'request to purchase' Kamaishi and a statement of intent, dated 4 March 1887, referred to the 'vast amounts of iron ore' (and charcoal) which were available

Table 5.2 Disposal of government enterprises

Enterprise	Government investment	Value of assets (June 1885)	Sale price	Percent of investment*	Percent of asset value*	Official sale date	Purchaser
Mines							
Takashima Coal Mines	393,848	—	550,000	139.65	—	November 1874	Gotō Shōjirō
Aburato Coal Mine	48,608	17,192	27,943	57.49	162.53	January 1884	Shirase
Nakakosaka Iron Mine	85,507	24,300	28,575	33.42	117.59	July 1884	Sakamoto Yuhachi
Kosaka Silver Mine	547,476	192,000	273,659	49.99	142.53	August 1884	Kuhara Shōzaburo
Innai Silver Mine	703,093	72,993	108,977	15.50	149.30	December 1884	Furukawa Ichibei
Ani Copper Mine	1,673,211	240,772	337,766	20.19	140.28	March 1885	Furukawa Ichibei
Okuzo & Magane Gold Mines	149,546	98,902	117,142	78.33	118.44	June 1885	Abe Sen (Hisomu)
Kamaishi Iron Mines	2,376,625	733,122	12,600	0.53	1.72	December 1887	Tanaka Chōbei
Miike Coal Mines	757,060	448,549	4,590,439	606.35	1023.40	December 1887	Sasaki Hachibei
Poronai Coal Mine & Railroad	2,291,500	—	352,318	15.37	—	November 1889	Hokkaido Tankō Tetsudō
Sado Gold Mine	1,419,244	445,250	2,560,926*	80.53	181.37	September 1896	Mitsubishi
Ikuno Silver Mine	1,760,866	966,752	—	—	—	September 1896	Mitsubishi
Cotton and Silk Mills							
Hiroshima Spinning Mill	54,205	—	12,570	23.19	—	June 1882	Hiroshima Menshi Bōseki
Aiichi Spinning Mill	58,000	—	—	—	—	November 1886	Mitsubishi
Shinmachi Spinning Mill	138,984	—	141,000	101.45	—	May 1887	Mitsui
Tomioka Filature	310,000	—	121,460	39.18	—	September 1893	Mitsui
Factories							
Fukugawa White Brick	93,276	—	83,862	89.91	—	July 1884	Asano Sōichirō
Shinagawa Glass	294,168	66,305	79,950	27.18	120.58	May 1885	Nishimura Katsuzō
Nagasaki Shipyard	1,130,949	459,000	459,000	40.59	100.00	July 1887	Mitsubishi
Hyogo Shipyard	816,139	320,196	188,029	23.04	58.72	March 1881	Kawasaki
Total	**15,102,305**		**10,046,216**				

Source: Based on Kobayashi, 1977: 138–9

Note: * Percentage of government investment/asset value represented in sale price. Total represents government investment and amount recovered by sale at the site. With these raw materials, Tanaka intended to resuscitate Japan's iron industry – as Matsukata wished – for the 'good of the nation.'[151]

With one move and a bureaucracy aligned behind him, Matsukata was able to dispose of Kamaishi, while still keeping it within reach. Best of all, his principle of directly supporting only agro-industries remained in tact. Matsukata believed that he had managed to ensure that at least some of Japan's iron needs would be met through domestic production, although this seemed a rather unrealistic assumption at the time.

Requiem to *bunmei kaika*

By 1895 and the end of the first Sino-Japanese War, *bunmei kaika* was dead. The word 'civilization' was still tossed around but all that it embodied was somehow different. The idea of attaining 'civilization' through Westernization was gone. Cultural materialists no longer promoted 'civilization and enlightenment.' 'Civilization' had been transformed into a level of achievement, a stage which marked the progress of a country's historical development in terms of its military and industry. The second generation of cultural materialist and progress ideologue described 'civilization' and 'progress' strictly in terms of Japanese interests. Iron machines, railroads, warships, and brick buildings continued to represent the higher levels of 'civilization' throughout the Victorian-era world; and Japan would continue to build these icons of 'civilization,' but not in an effort to become Westernized. The new expression of Meiji identity was more or less complete, not only in terms of a particularistic Japanese formula, but also in terms of the iron and iron artifacts that would help to define the Japanese nation.[152]

The technological artifacts so thoroughly infused into the movement to attain 'civilization and enlightenment' also contributed to the movement's demise. Icons of the West served as the cultural materialist's proof that Japan was attaining a higher 'level of civilization' while simultaneously providing fuel for the critics' claims of hollow mimicry. One needed look no further than the daily press to find a continuous barrage of comments attacking the government's efforts to increase Japan's 'level of civilization.' From the time the Iwakura Mission left for Europe and the United States, foreign critics discussed the futility of Japanese efforts to 'attain civilization.' Charges that the Japanese did not understand the most fundamental principles of the Enlightenment accompanied nearly every positive statement of the island nation's technological progress.[153] Many Japanese too, were critical of their government's efforts, as people at all levels of society began to abandon the icons of Western civilization in favor of more indigenous alternatives.[154]

Kawase Hideharu, Senior Deputy Minister (*daijō*) of the Ministry of Home Affairs and Deputy Chief (*gonnokami*) of the Industrial Promotion Bureau (*Kangyō kyoku*), summed up the critics' position well. He argued that the 'government's progress toward civilization was ill conceived. It was based on the will of people whose decisions are above logic.' In short, he stated, 'Our country's progress toward civilization is not true progress at all.'[155] Kawase was referring to government policies that promoted the adoption of Western technological artifacts and the inability of the people to benefit from the new technologies which they struggled to utilize.[156] According to the deputy minister, the government was attempting to change the people's spirit, a task which could not be accomplished in such a short time. Official enthusiasm for *materia Hesperia* only contributed to the problem. 'Enlightenment promoters,' overly-enthusiastic after visiting Europe and the United States, compounded the issue because they too, called for rapid change. The Meiji government could not simultaneously awaken the spirit of industry in the people, change their values, overhaul industry, and turn Westerners into consumers of Japan's new industrial products in less than a decade.[157]

Dichotomies inherent in the so-called program of industrialization, *shokusan kōgyō*, were part of the problem. Meiji officials and intellectuals were simultaneously promoting Western technological artifacts as the means by which Japan would become 'civilized,' remain independent of the West, and build a powerful military. Concurrently, however, was the necessity of abandoning Japan's history and promoting liberal theories which had little meaning to the majority of Japanese society – and were applied with greater frugality to Japan by the West. The result was a natural contradiction, expressed perhaps best by Fukuzawa Yukichi, who had gone from being an internalist promoter of 'enlightenment thought' to one who promoted the cause of arms in the name of Japanese independence and regional hegemony.[158]

Japan was undergoing a thorough intellectual crisis. As Kenneth Pyle has aptly demonstrated, Meiji identity was in flux. Cultural materialists first embraced Western technical achievement and technological artifacts without reservation to begin their program of 'civilization building.' They constructed Japan's identity in terms of a new set of cultural icons. Iron bridges, brick buildings, railroads, telegraphs, iron machines, clothing, food, and even haircuts would help define Japan's identity for the first decade of Meiji. Contradictions between the idealized Western civilization and reality struck hard and fast, however, and people soon began to question a policy which sought to define Japan in European or American terms. Within the identity crisis which occupied the greater part of Meiji intellectual discourse were situated political and economic crises, which also helped define the 'new' Japanese identity.

When Matsukata came to power in mid-October 1881, he was able to promote his own agenda in the name of financial retrenchment. Few were willing to question his motives or actions, or challenge the newly re-invigorated Sat-Chō alliance. Even Fukuzawa, an Ōkuma supporter, argued that the government needed to become more unified, and sacrificed his support of the People's Rights Movement in favor of a more conservative 'anti-Western' stance.[159] The Meiji bureaucracy also saw political challenges to its *techno-diplomatic* identity from China and Korea. While the Qing were in the midst of their own modernization program, the 'hermit kingdom' of Korea rejected Japan's attempts to export 'civilization and enlighten-ment.' The West was embarking on its own renewed surge of imperialism and, if not openly challenging Japanese political interests, was denying the island nation credit for all the 'progress' it had made.[160]

As intellectuals struggled with their own new identities, Meiji officials began similarly to recast the nation's identity in terms of a new set of technological artifacts. Still made from modern materials, that is iron and steel, the icons of the new Japan, warships and heavy guns, simultaneously expressed modernity and Japanese-ness. Tools of war were not inherently Japanese, but they represented renewed priorities of 'national security' and 'military preparedness,' and the ability of the Meiji government to express its political will in terms of what Clausewitz called 'the continuation of policy by other means' – war.[161] The consequences of this new Meiji identity would become fully apparent in 1894.

6 Conclusion

From technological determinism to techno-imperialism

Japan in the mid-1890s was a very different place from the one that Perry encountered a few decades before. The technological landscape had changed along with the way that people thought about technological artifacts indigenous and foreign. Initially indigenous technologies were considered an acceptable base upon which to build the nation. Built from wood, however, Japan's machines, like its buildings, would not stand the test of 'civilization' as foreign beliefs in the superiority of iron and brick forged in Japan a progress ideology of materials. Within a few short years of the restoration of imperial rule, iron girders replaced the wooden beams which spanned many of Japan's rivers, brick buildings stood where wooden structures once endured, and iron machines supplanted their functional wooden equivalents.

Japan's technological transformation was not happenstance. It grew out of the political, social, and economic loci of Meiji modernization efforts. Meiji officials were faced with a daunting task, one for which they had vision but no single plan. They needed to unify the country – raising its 'level of civilization' – while demonstrating the legitimacy of their new order to the former domain authorities. Simultaneously, they had to provide similar evidence to the foreign powers whose presence in the recently established treaty ports was a constant reminder that Japan too, could share the ignominious fate of Qing China. Meiji officials, like the many samurai from whom they became distanced, knew the importance of 'knowing one's adversary.'[1] Early contacts with the West provided the basis of Meiji cultural materialism as men like Shibusawa Eiichi, Sugiura Yuzuru, Fukuzawa Yukichi, and later Itō Hirobumi and Ōkubo Toshimichi quickly learned that technological artifacts produced from iron and powered by steam were as crucial for measuring a country's physical 'level of civilization' as they were for the ideological super-structure of 'civilization building.' Each trip to the West only served to reinforce what already seemed apparent: Europe and the United States represented higher 'levels of civilization' as exemplified by their dominant positions in the world, made possible by demonstrably superior technologies.[2] A 'semi-civilized' nation such as Japan would have to rapidly assimilate Western technology in order to survive. Originally embraced as the method by which to raise Japan's 'level of civilization,' icons of Western material culture were quickly embedded within the Japanese psyche as evidence that the newly emerging nation was already becoming more 'civilized' than its Asian neighbors as it embarked on a trajectory to achieve equality with the West.

Japan's early Meiji embrace of Western material culture is widely recognized for its ties with the so-called movement to attain 'civilization and enlightenment.' Historians, not to mention contemporary critics, have regularly targeted the public's embrace of what were considered superficial elements of the West, such as clothing, umbrellas, pocket-watches, and hairstyles, as evidence of the desultory nature of *bunmei kaika* materiality. The government's externalist role, that is the theoretically unprincipled modernization of Japan through the absorption of Western technological artifacts, is also interpreted, at least in part, as a component of *bunmei kaika*. But the potential of cultural materiality and its transformative ability for both artifact and society has not heretofore been recognized.

Through their promotion and the absorption of Western technological artifacts, would-be Meiji modernizers began a process of embedding a new set of cultural values in Japanese society. With each artifact came some symbolic attachment. Iron artifacts among other things, represented precision and 'civilized' man's control over his natural environment. The emperor's new Western-style military uniform elevated him to the unfamiliar role of supreme military commander, while simultaneously distancing Japan from its past attachments to Chinese culture; brick ministerial buildings represented permanence, Western-style diplomacy, and concurrently a higher 'level of civilization.' While some of Meiji Japan's new cultural values were relatively insignificant for many, others were becoming more troublesome by the 1890s because of the degree to which they denied Japan's past.

Artifacts which had for centuries been part of Japanese material culture were suddenly divorced from their technological and psychological landscapes. While the emperor's new clothes served to distance a 'civilized' Japan from an 'uncivilized' China, they simultaneously placed a similar label on much of Japan's history and culture. Phrases such as 'evil customs of the past' were tossed around with a deliberate ambiguity that allowed them to be applied with the broadest of strokes.[3] For externalists seeking to import the icons of Western civilization as instruments of treaty revision, this was not a concern. The essence of Japanese culture was not being questioned, only its physical manifestations. Everything from buildings to rickshaw and silk reeling machinery to ships had to reflect a new 'level of civilization' that would earn Japan the foreign powers' respect.[4]

Internalists, on the other hand, believed that there was something in Japan's past that had delayed its 'progress toward civilization,' thus keeping Japan from taking its proper place amongst the 'civilized' nations of the world. Japanese thought needed to be infused with the spirit of the European Enlightenment and scientific inquiry. Only then would Japan once again be set on its proper course where freedom of action would 'elevate the human spirit' and make beneficial the absorption of Western technological artifacts.[5] Largely from the old tradition, externalist and internalist alike found little problem in restructuring 'modern' Japan in terms of Western technological artifacts, foreign knowledge, or 'enlightenment' thought.[6] To Meiji officials and intellectuals, importing Western civilization, whether technological artifact or ideological construct, was a matter of national prestige and raising

Japan's 'level of civilization.' Neither party intended to destroy the basis of Japan's past; in fact, they used its history to strengthen the nation.[7]

Artifacts like ideologies have a fluid existence in the socio-cultural continuum. Just as the 'new generation'of Meiji intellectual struggled with an identity in flux 'between the familiar, comfortable values of their [Confucian-based] childhood and the new, liberating values of their later [Western] education,' the technological artifacts that were once symbolic of 'civilization and enlightenment' gradually assumed new cultural and political meanings.[8] The artifact's ability to transform nineteenth-century Japan into a 'modern' nation was reflected as the new Meiji culture's ability to impart a different set of values to these same technological artifacts or the materials of their construction. Through their progress ideology of materials, government official and entrepreneur alike quickly elevated Western iron artifacts to the highest levels of 'civilization and enlightenment.' The externalist's iron machines and other artifacts seemingly infused Meiji culture with the internalist's 'enlightenment' principles. But intercourse with the West illustrated the fleeting nature of the 'enlightened spirit' as official and intellectual quickly learned that the West's progressive values did not necessarily apply to all.[9] Diplomacy in an era of high imperialism only served to heighten Japan's sense of isolation and perceived need for a *strong army*. Iron, the material of 'civilization,' was transformed into the material of national defense and military preparedness by officials and engineers who renewed their call for a modern, national iron industry as the basis of Japan's modern military.

Despite internal opposition, Japan largely relied on imported iron until the end of the nineteenth century while it pursued a policy of *techno-imperialism*.[10] Only after their victory in the first Sino-Japanese War was overshadowed by the humiliation of the Triple Intervention, was the urgency of building a national iron industry pushed to the fore. Meiji officials finally 'understood' that Japan could not rely on foreign iron as the foundation for either a *strong army* or *rich nation*. In concert with a host of other factors, Japan's modern navy's insatiable appetite for the former material of 'civilization and enlightenment' would set Japan on an eventual collision course with the West.[11]

The technological transformation of Japan is frequently discussed in terms of the relationship between political need and the economic implications of government initiatives, however ill-conceived, which laid the foundations of modern Japan's industrial development.[12] Some of the more recent scholarship focuses on the relationship between the local, small-scale producer and the government's enterprises.[13] And while most of these studies recognize the role which 'civilization and enlightenment' thinking played in both popular and official acceptance of Western technological artifacts, none has recognized the existence or pervasiveness of a progress ideology of materials and its effect on choice of technique. Meiji officials rejected viable technologies because they lacked symbolism – a seemingly irrational choice. Believing wooden machines were too traditional and therefore inferior, they embraced cast iron as a thoroughly 'modern' material from which to construct 'modern' industries and by extension, a 'modern and civilized' nation. So strong was their ideological perspective that officials ignored the fact that their choices were frequently neither economically nor technologically rational. The choice of French cast-iron machinery and the desire to use imported French coal at Tomioka Silk Filature are perfect examples. Understanding choice of technique and labeling it rational or irrational, however, requires that we identify the purpose, or purposes, for which the technology was selected.

For would-be Meiji modernizers, material symbolism was equally, if not more important, than the economic or technical feasibility of any given technological artifact. In the rush to bolster their all but secure position, Meiji officials of the early-1870s followed a policy of *techno-diplomacy* in choosing multi-functional technological artifacts: ones that they believed would simultaneously build the economy, promote political legitimacy, and raise the newly forming nation's 'level of civilization.' As these same officials became more assured of their authority, domestically and internationally, they began to retreat from a posture of seemingly unquestioned acceptance of Western material culture toward a new policy which I call *selective techno-political restructuring*. Still believing that some icons of Western 'civilization' would raise Japan's status among the nations of the world, Meiji officials and other public figures continued to embrace certain artifacts, such as railroads, steam engines, and heavy ordnance, while simultaneously rejecting others.[14]

Despite what has often been categorized as blind acceptance of the West, symbolized perhaps best by the *Rokumeikan* (a building designed by Josiah Conder for entertaining foreign visitors), many Japanese had for nearly a decade been moving away from this type of thinking. *Bunmei kaika* was a 'monster' (*kaibutsu*) to internalist social critics like Kawamura Yoshia who argued in 1877 that Japan needed to 'choose wisely' when accepting the artifacts of Western 'civilization.'[15] Tokutomi Sohō's identity crisis of the 1890s was an earlier dilemma for men like Fukuzawa Yukichi who vacillated between internalist and externalist, liberal and conservative as they struggled to find a socio-political middle ground between Western iconic acceptance and Japanese cultural independence.[16]

In the name of fiscal retrenchment, the Meiji government began to abandon many of the very enterprises which had brought the nation international prestige and recognition. Having faithfully served to raise Japan's 'level of civilization,' and after almost a decade of debate, the star of Meiji *techno-civilization*, Tomioka Silk Filature, was quietly sold to Mitsui. The Shinmachi Waste Silk factory, upon which many of the same hopes were pinned, shared a similar fate.[17] Despite a growing recognition that Japan needed a viable, modern iron industry and inter-ministerial support for such a project, Kamaishi Ironworks was also sold at this time. Typically considered 'the most eligible candidate' for disposal among the government's supposed entrepreneurial failures, Kamaishi's sale is considered only in terms of economics.[18] The government's ironworks, however, lacked the symbolic importance of Tomioka. Unlike the case of its premier filature, neither Britain nor France shared any interest in Kamaishi. And while the foreign powers had some early hopes for the government's ironworks, their inability to participate in the extraction of Japan's mineral wealth meant that Kamaishi would be built in relative isolation.[19]

For a government that struggled for foreign recognition of its efforts to raise Japan's 'level of civilization,' the inability to keep Kamaishi in-blast was perceived as a step backward for national prestige. Not recognizing the relatively imprecise nature of iron smelting, officials like Finance Minister Matsukata, who were also ideologically opposed to publically funding large Western-style industries, were quick to label the ironworks a failure. Although not sharing the slide into oblivion that was to become the fate of the Ministry of Public Works' Akasaka filature, Kamaishi was ingloriously disposed of under highly questionable circumstances nonetheless. In much the same way that Meiji officials began their quest to embed the technological artifacts of Western civilization into the Japanese cultural landscape – with multiple, often unstated purposes – a similar group of officials disposed of these same artifacts for just as many unstated reasons.

In retrospect, the Ministry of Public Works' inability to keep Kamaishi in-blast illustrates the fallacy of early Meiji beliefs in technological determinism. 'Universal' iron-smelting technologies were not 'universal'; progress was not linear. Long before the government sold Kamaishi, however, cultural materialists had unknowingly begun their interactive transformation of Meiji culture and technological artifacts. Japanese society was not simply adapting to new technologies. Artifactual icons of the West were used to define Japan's 'level of civilization' as cultural materialists simultaneously defined these same artifacts in terms of politico-cultural values. Cast-iron machinery, described as 'sophisticated,' 'civilized,' and 'enlightened,' was embedded in Japan as cultural materialists sought to infuse these values into society. As the Japanese became more 'civilized and enlightened,' furthering their understanding of the West, however, iron's politico-cultural values changed as this material of 'progress' was transformed into the materiel of military preparedness, national defense, and war. The same progress ideology of materials which allowed Meiji Japan to embrace iron as the material from which to build 'civilization,' helped redefine iron in terms of the nation's newest form of diplomatic expression, imperialism.

Appendix I
Proposals for the location of Kamaishi Ironworks[1]

The following are translations of Louis Bianchi's and Ōshima Takato's 1874 proposals for the location of Kamaishi Ironworks. Because many accounts of the development of Japan's iron industry focus on these proposals, or the repercussions of the government having chosen Bianchi's plan, I have included the texts. Bianchi's proposal is named *Suzukosetsu* for the site he selected, Suzuko, to the south east of Kamaishi village. Ōshima's proposal, Otadagoesetsu, is named for its site, *Otadagoe*, on Kamaishi's north east side.

Suzukosetsu

1. There is easy access to water, which is only 1,900 *shaku* (57 meters) from the site. Even though we can get water from the river, the construction site should be close because we do not need the added expense.
2. Since the parcel of land is very wide, there will be no need to expand the site when the time comes to expand the facility. At such a time, there will also be the need for unmeasurable quantities of water.
3. If the side of the mountain is utilized when building the chimneys, roads, and storage facilities for raw materials, costs and building materials will be economized.
4. There is sufficient vacant land [at this location] to dispose of slag.
5. It will be easy to bring either charcoal to the ironworks from the harbor, or iron to the harbor from the ironworks if we use a dock and/or railroad line.
6. Stone for construction of the earthen works is immediately abundant on the land.
7. Compared to *Otadagoe*, the distance from Kamaishi Bay to the ironworks is no different. Even if the land [at Otadagoe] is suitable, it will cost less to build on the mountainside and be more expensive in the valleys and flat land. Moreover, labor costs and construction costs for the necessary facilities will be lower.[2]

Otadagoesetsu

1. The topography of *Otadagoe* to the east makes it the best choice. It is surrounded by mountains on three sides: north, west, and east, only the south is open. Throughout the four seasons there are no rainstorms, and even in severe winters, the cold will not stop production day or night.

2. If we flatten the top of the mountain and use it as a place for storing iron ore, charcoal, etc., and build the furnace and install the machinery at the base of the mountain, there is plenty of space to work.

3. By utilizing this site, there will be no fear of blocking public roads while transporting ore from the mine to the iron works. Moreover, because railroads are very expensive, if we use 70 horses and a tramway, it will be possible to transport the ore necessary for (only) five furnaces. In addition, if we construct a cable car or tramway there will be no need to worry about blocking public roads [with railroad tracks].

4. Regarding water for machinery, although it may be one *ri* (3.9 kilometers) away, we can raise water from upstream and bring it to the site by running iron pipes around the side of the mountain.

5. Transporting charcoal and iron will not be a problem if a canal (channel) is dug from the iron works to the harbor. To dig a new canal would not be more than 1950 *shaku* (477 meters).

Appendix II

Distribution and productivity
of filatures by prefecture

The following shows the distribution and productivity of filatures, grouped by prefecture, as part of an 1895 national survey. Number of basins in column C equals the number of reeling frames. Columns F and G represent the prefecture's percentage of national production by machine type. 'Mechanized' refers to either Western-style reeling frames or hybrid machines which are distinguished from traditional *zaguri* machines.[1]

National silk reeling productivity (filatures with more than 10 employees)

	A	B	C	D	E	F	G
	Type of filature	Number of filatures	Number of basins	output (in kin/ year)[2]	% total filatures	% total basins	% total production
Tokyo-fu	zaguri	3	96	50,106	20.00	0.5	11.86
	mechanized	12	1,214	84,960	80.00	1.41	2.53
	total	*15*	*1,310*	*135,066*	*0.46*	*1.25*	*3.57*
Kyoto-fu	zaguri	10	131	2,325	4.38	0.69	0.05
	mechanized	218	3,618	77,726	95.61	4.2	2.32
	total	*228*	*3,749*	*80,851*	*7.13*	*3.57*	*2.13*
Osaka-fu	zaguri	7	134	1,227	53.84	0.7	0.29
	mechanized	6	203	6,407	46.15	0.23	0.19
	total	*13*	*337*	*7,634*	*0.40*	*0.32*	*0.2*
Kanagawa	zaguri	4	236	4,600	20.00	1.23	1.08
	mechanized	16	912	25,639	80.00	1.06	0.76
	total	*20*	*1,148*	*30,239*	*0.62*	*1.09*	*0.8*
Hyogo	zaguri	28	385	5,455	30.10	2.013	1.29
	mechanized	65	1,977	55,101	69.89	2.3	1.64
	total	*93*	*2,362*	*60,556*	*2.91*	*2.25*	*1.6*
Nagasaki	zaguri	4	124	1,098	66.66	0.65	0.26
	mechanized	2	100	3,800	33.33	0.11	0.11
	total	*6*	*224*	*4,898*	*0.19*	*0.21*	*0.13*
Niigata	zaguri	26	531	6,339	44.06	2.78	1.5
	mechanized	33	1,137	24,291	55.93	1.32	0.72
	total	*59*	*1,668*	*30,630*	*1.84*	*1.59*	*0.81*
Saitama	zaguri	32	2,040	65,485	78.04	10.66	15.5
	mechanized	9	807	43,363	21.95	0.94	1.29
	total	*41*	*2,847*	*108,848*	*1.28*	*2.7*	*2.88*

	A	B	C	D	E	F	G
	Type of filature	Number of filatures	Number of basins	output (in kin/year)²	% total filatures	% total basins	% total production
Gunma	zaguri	48	1,071	27,027	52.74	5.6	6.4
	mechanized	43	1,968	80,626	47.25	2.29	2.4
	total	*91*	*2,985*	*107,653*	*2.84*	*2.84*	*2.84*
Chiba	zaguri	18	581	13,463	51.42	3.03	3.18
	mechanized	17	596	18,704	48.57	0.69	0.56
	total	*35*	*1,177*	*32,167*	*1.09*	*1.12*	*0.85*
Ibaragi	zaguri	95	1,877	49,078	71.96	9.8	11.6
	mechanized	37	1,793	68,891	28.03	2.08	2.05
	total	*132*	*3,670*	*117,969*	*4.12*	*3.5*	*3.04*
Tochigi	zaguri	30	652	21,304	96.77	3.4	5.04
	mechanized	1	200	17,500	3.22	0.23	0.52
	total	*31*	*852*	*38,804*	*0.97*	*0.81*	*1.02*
Mie	zaguri	17	239	4,381	20.98	1.24	9.18
	mechanized	64	1,346	60,961	79.01	1.57	1.82
	total	*81*	*1,585*	*65,342*	*2.53*	*1.5*	*1.73*
Aichi	zaguri	15	325	3,921	7.57	1.7	0.93
	mechanized	183	4,974	179,708	92.42	5.79	5.36
	total	*198*	*5,299*	*183,629*	*6.19*	*5.04*	*4.86*
Shizuoka	zaguri	10	170	6,584	19.60	0.89	1.56
	mechanized	41	1,600	91,752	80.39	1.86	2.73
	total	*51*	*1,770*	*98,336*	*1.59*	*1.68*	*2.6*
Yamanashi	zaguri	7	92	1,381	2.78	0.48	0.33
	mechanized	244	8,929	251,565	97.21	10.39	7.5
	total	*251*	*9,021*	*252,946*	*7.84*	*8.58*	*6.69*
Shiga	zaguri	5	58	1,044	12.50	0.3	0.25
	mechanized	35	1,509	59,321	87.50	1.76	1.77
	total	*40*	*1,567*	*60,425*	*1.25*	*1.49*	*1.6*
Gifu	zaguri	4	44	2,126	0.80	0.23	0.5
	mechanized	493	10,497	267,449	99.19	12.26	7.97
	total	*497*	*10,541*	*269,575*	*15.54*	*10.03*	*7.13*
Nagano	zaguri	14	1,293	40,121	2.72	6.76	9.5
	mechanized	500	24,869	1,335,221	97.27	28.93	39.79
	total	*514*	*26,162*	*1,375,342*	*16.07*	*24.9*	*36.4*
Miyagi	zaguri	19	683	7,982	52.77	3.57	1.89
	mechanized	17	1,052	34,219	47.22	1.22	1.02
	total	*36*	*1,735*	*42,201*	*1.13*	*1.65*	*1.12*
Fukushima	zaguri	14	3,976	40,995	35.00	20.79	9.7
	mechanized	26	1,222	38,343	65.00	1.42	1.14
	total	*40*	*5,198*	*79,338*	*1.25*	*4.95*	*2.1*
Iwate	zaguri	9	395	1,703	23.68	2.065	0.4
	mechanized	29	1,022	16,470	76.31	1.18	0.49
	total	*38*	*1,417*	*18,173*	*1.18*	*1.35*	*0.48*
Aomori	zaguri	2	143	1,663	25.00	0.75	0.39
	mechanized	6	72	805	75.00	0.083	0.023
	total	*8*	*217*	*2,468*	*0.25*	*0.21*	*0.065*
Yamagata	zaguri	0	0	0	0.00	0	0
	mechanized	109	2,709	99,687	100.00	3.15	2.97
	total	*109*	*2,709*	*99,687*	*3.40*	*2.58*	*2.63*

	A	B	C	D	E	F	G
	Type of filature	Number of filatures	Number of basins	output (in kin/year)[2]	% total filatures	% total basins	% total production
Akita	zaguri	5	64	559	100.00	0.026	0.14
	mechanized	0	0	0	0.00	0	0
	total	*5*	*64*	*559*	*0.15*	*0.06*	*0.015*
Fukui	zaguri	10	170	7,159	30.30	0.89	1.69
	mechanized	23	1,000	53,673	69.69	1.16	1.6
	total	*33*	*1,170*	*60,832*	*1.03*	*1.11*	*1.61*
Ishikawa	zaguri	19	234	5,651	20.65	1.22	1.34
	mechanized	73	1,196	33,224	79.34	1.39	0.99
	total	*92*	*1,430*	*38,875*	*2.88*	*1.36*	*1.02*
Toyama	zaguri	0	0	0	0.00	0	0
	mechanized	97	2,459	77,676	100.00	2.86	2.3
	total	*97*	*2,459*	*77,676*	*3.03*	*2.34*	*2.05*
Tottori	zaguri	0	0	0	0.00	0	0
	mechanized	61	1,630	59,683	100.00	1.9	1.78
	total	*61*	*1,630*	*59,683*	*1.91*	*1.55*	*1.58*
Shimane	zaguri	41	1,275	13,884	68.33	6.67	3.29
	mechanized	19	804	32,638	31.66	0.94	0.97
	total	*60*	*2,079*	*46,522*	*1.88*	*1.98*	*1.23*
Okayama	zaguri	16	286	4,468	32.65	1.49	1.06
	mechanized	33	1,121	44,009	67.34	1.3	1.31
	total	*49*	*1,430*	*48,477*	*1.53*	*1.36*	*1.28*
Hiroshima	zaguri	7	135	4,751	19.45	0.7	1.12
	mechanized	8	559	20,807	80.54	0.65	0.62
	total	*15*	*694*	*25,558*	*0.46*	*0.66*	*0.67*
Yamaguchi	zaguri	0	0	0	0.00	0	0
	mechanized	7	403	13,951	100.00	0.47	0.42
	total	*7*	*403*	*13,951*	*0.21*	*0.38*	*0.37*
Wakayama	zaguri	1	14	163	3.13	0.073	0.04
	mechanized	14	432	20,306	97.73	0.5	0.61
	total	*15*	*446*	*20,469*	*0.46*	*0.42*	*0.54*
Tokushima	zaguri	0	0	0	0.00	0	0
	mechanized	4	129	6,094	100.00	0.15	0.18
	total	*4*	*129*	*6,094*	*0.13*	*0.12*	*0.16*
Kagawa	zaguri	0	0	0	0.00	0	0
	mechanized	4	105	4,090	100.00	0.122	0.12
	total	*4*	*105*	*4,090*	*0.13*	*0.099*	*0.11*
Ehime	zaguri	7	85	791	13.23	0.44	0.19
	mechanized	15	557	2,326	86.76	0.65	0.069
	total	*22*	*642*	*23,117*	*0.69*	*0.61*	*0.61*
Kochi	zaguri	15	387	7,040	100.00	2.02	1.67
	mechanized	0	0	0	0.00	0	0
	total	*15*	*387*	*7,040*	*0.47*	*0.37*	*0.19*
Fukuoka	zaguri	11	275	2,824	44.57	1.44	0.68
	mechanized	6	352	11,869	57.05	0.41	0.35
	total	*17*	*617*	*14,693*	*0.53*	*0.59*	*0.39*
Oita	zaguri	5	68	1,769	12.03	0.36	0.42
	mechanized	23	499	17,064	88.31	0.58	0.51
	total	*28*	*565*	*18,833*	*0.88*	*0.54*	*0.5*

	A	B	C	D	E	F	G
	Type of filature	Number of filatures	Number of basins	output (in kin/ year)^2	% total filatures	% total basins	% total production
Kumamoto	zaguri	5	75	1,009	60.00	0.39	0.24
	mechanized	1	50	2,800	40.00	0.058	0.083
	total	*6*	*125*	*3,809*	*0.19*	*0.12*	*0.1*
Miyazaki	zaguri	4	92	1,407	30.26	0.48	0.33
	mechanized	5	212	9,162	69.73	0.25	0.27
	total	*9*	*304*	*10,569*	*0.28*	*0.29*	*0.28*
Kagoshima	zaguri	25	643	9,291	91.72	3.36	2.2
	mechanized	3	58	669	8.27	0.067	0.019
	total	*28*	*701*	*9,960*	*0.88*	*0.67*	*0.26*
Hokkaido	zaguri	2	45	2,207	41.28	0.24	0.52
	mechanized	2	64	3,281	58.71	0.074	0.097
	total	*4*	*109*	*5,488*	*1.30*	*0.1*	*0.15*
National total							
	zaguri	594	19,124	422,381	18.57	18.12	11.18
	mechanized	2,604	85,958	3,355,831	81.42	81.8	88.82
	total	3,198	105,082	3,778,212			

Notes

1 Introduction

1 The phrase 'Japanese enlightenment' is based on Carmen Blacker's study of Fukuzawa Yukichi. Carmen Blacker, *The Japanese Enlightenment: A Study of the Writings of Fukuzawa Yukichi*, Cambridge: Cambridge University Press, 1964.

2 For example see: Yonekura Seiichiro, *The Japanese Iron and Steel Industry, 1850–1990: Continuity and Discontinuity*, New York: St. Martin's Press, 1994; Iida Kenichi, *Gendai nihon no gijutsu to shisō*, Tokyo: Tōyō Keizai Shinpōsha, 1974; Iida Kenichi, *Nihon tekkō gijutsushiron*, Tokyo: Mihari Shobō, 1973; Hanzawa Shūzō, *Nihon seitetsu kotohajime: Ōshima Takatō no shōgai*, Tokyo: Shinjinbutsu Oraisha, 1974; Andō Yasuo, 'Tomioka seishijo,' Chihōshi Kenkyū Kyōgikai, ed., *Nihon sangyōshi taikei*, vol. 4, Tokyo: Tokyo Daigaku Shuppankai, 1959; and Kamijō Hiroyuki, *Kinu hitotsuji no seishun: 'Tomioka nikki' ni miru Nihon no kindai*, Tokyo: NHK Books, 1978.

3 There are a number of studies which attempt to put the Japanese experience into its international context without turning Japan into a developmental model. See Claudio Zanier, 'Japan and the 'Pébrine' Crisis of European Sericulture During the 1860s,' in Erich Pauer, ed., *Silkworms, Oil, and Chips . . .*, Bonn: Japanologisches Seminar, 1986; Sugiyama Shinya, *Japan's Industrialization in the World Economy 1859–1899: Export Trade and Overseas Competition*, London: Anthlone Press, 1988; William D. Wray, ed., *Managing Industrial Enterprise: Cases from Japan's Prewar Experience*, Cambridge, MA: Harvard University Press, 1989. The last volume, although not specifically aimed at putting Japan's industrialization efforts into an international context, places them within the larger scope of Japanese history.

4 For examples of works using Japan's historical industrialization as a developmental model for NIEs, see: Hayashi Takeshi, *The Japanese Experience in Technology: From Transfer to Self-Reliance*, Tokyo: United Nations University Press, 1990; Minami, Ryōshin, Kwan S. Kim, Fumio Makino, and Joung–hae Seo, eds., *Acquiring, Adapting and Developing Technologies: Lessons from the Japanese Experience*, New York: St. Martin's Press, 1995; Yoshihara Kunio, *Japanese Economic Development*, Oxford: Oxford University Press, 1994; and Gotō Akira and Hiroyuki Odagiri, eds, *Innovation in Japan*, Oxford: Clarendon Press, 1997.

5 See for example, Walter G. Vincenti, *What Engineers Know and How They Know It: Analytical Studies from Aeronautical History*, Baltimore, MD: Johns Hopkins University Press, 1990; and Thomas S. Kuhn, *The Structure of Scientific Revolutions*, Chicago, IL: University of Chicago Press, 1970.

6 Debin Ma, 'The Modern Silk Road: The Global Raw-Silk Market, 1850–1930,' *The Journal of Economic History*, 56: 2, (June, 1996), p. 341.

7 Sano Tsunetami, *Ōkoku hakurankai hōkokusho: sangyōbu*, vol. 1, section 1: 7, Tokyo, 1875. (Hereafter, *OHHS*).

8 See Minbushō directive of 1870 reprinted in Ōtsuka Ryōtarō, *Sanshi*, 2 vols. Tokyo: Fusōen, 1900, 1: 249–50.

9 McCallion rightly argues that there were conflicting functions for Tomioka. Publicly, the government stated that the facility was a model and men like Shibusawa Eiichi maintained this position well into the twentieth century. At other times Tomioka was considered an enterprise, and profit-making became the primary consideration. These often conflicting goals and the government's inexperience in operating a business contributed to many of Tomioka's difficulties. Stephen McCallion, *Silk Reeling in Japan: The Limits to Change*, Ph.D. dissertation, Columbus, OH: The Ohio State University, 1983. This work is probably the most comprehensive English-language examination of the development of Japan's silk reeling industry available.

10 *Kōbushō* is also translated as Ministry of Industry, Ministry of Engineering, Ministry of Construction, and Ministry of Technology. Because many of its programs were either designed to serve the public good or were public in nature, I have chosen to use Ministry of Public Works.

11 Yamao Yōzō, 'Kōbushō kōzan kaigyō no kyūmu taru o jūzosu,' *Dajō ruiten*, 2: 118, Chihō 24: 1, 4 January 1872.

12 Japan's first mining laws were based on European law. For a description of the various mining laws see, Fukui Jun, *Nihon kōhō zenshū*, Osaka: Hōgyokudō, 1887. For a rough English translation of early Meiji mining laws see: Curt Netto, *Memoirs of the Science Department, University of Tokio, Japan*, vol. 2, 'On Mining and Mines in Japan,' Tokyo: University of Tokio, 1879, pp. 22–3.

13 Robert Friedel, 'Some Matter of Substance' in Steven Lubar and W. David Kingery, eds, *History from Things: Essays on Material Culture*, Washington, DC: Smithsonian Institution Press, 1993, p. 42.

14 See Kannōkyoku, Shomukyoku, *Kyōshinkai hōkoku, kenshi no bu*, Tokyo: Yūrindō, 1880, pp. 24–89 *passim* (hereafter *KH*). The Tomioka mystique and frequency of false or exaggerated claims will be discussed in greater detail in Chapter 3.

15 See Jacques Maquet, 'Objects as Instruments and Signs' in Lubar and Kingery, *History from Things*, and Paul Ricoeur, Kathleen Blamey and John B. Thompson, trans., *From Text into Action*, Evanston, IL: Northwestern University Press, 1991.

16 Ricoeur makes this claim for texts, i.e., written words have independence from the author once transmitted because they are interpreted differently by anyone who reads them: Ricoeur, *From Text into Action*, p. 158.

17 Jules David Prown, 'The Truth of Material Culture: History or Fiction?' in Lubar and Kingery, *History from Things*, p. 1.

18 The phrase 'Victorian Japan' is from Dallas Finn, *Meiji Revisited: Sites of Victorian Japan*, New York: Weatherhill, 1995, p. 3.

19 In traditional major silk-producing regions such as Gunma and Fukushima prefectures, diffusion of Western-style mechanization took until around 1910. In Fukushima for example 95% of the raw silk was produced by traditional methods as late as 1895. Despite being the home to Tomioka Silk Filature, Gunma prefecture was equally as slow; its silk exports largely relied on machinery of improved traditional methods. Non-traditional silk producing regions that saw the opportunity to move into the lucrative market, such as Nagano prefecture, were mechanized based on Western methods by the 1880s. See Sugiyama Shinya, *Japan's Industrialization in the World Economy 1859–1899: Export Trade and Overseas Competition*, London: Anthlone Press, 1988, pp. 112–128; and Nōshōmushō Nōmukyoku, *Daini seishikōjo chōsahyō*, Tokyo: Nōshōmushō Nōmukyoku, 1898, pp. 35–9 (Gunma), 120–46 (Nagano), 150–8 (Fukushima).

20 For discussions of various deterministic theories of technology see, Andrew Feenberg and Alastair Hannay, *Technology and the Politics of Knowledge*, Bloomington, IN: Indiana University Press, 1995; and Merritt Roe Smith, *Does Technology Drive History? The Dilemma of Technological Determinism*, Cambridge: MIT Press, 1994.

21 Eric Schatzberg, *Wings of Wood, Wings of Metal: Culture and Technical Change in American Airplane Materials, 1914–1945*, Princeton, PA: Princeton University Press, 1999, pp. 3–4.
22 America, a wooden country, was of course the exception and a point of mild controversy for British architects in Meiji Japan. Most attributed America's wooden cities to the pioneer spirit that was gradually being swept away by the more permanent stone structures that were coming to line American city streets; see Gregory K. Clancey, *Earthquake Nation: The Cultural Politics of Japanese Seismology, 1868–1930*, Berkeley, CA: University of California Press, 2005, p. 15, n. 22.
23 Ōtsuka Ryōtarō, *Sanshi*, 2 vols. Tokyo: Fusōen, 1900, 1: 249–50.
24 'Rikukaigun kōbu sanshō gōgi shi ichidai seitetsujo o sōken sen to su jōte gaikokujin o yatoire,' *Dajō ruiten*, (gaikoku kokusai), 2: 72, no. 36 (4 May 1875).
25 'Kōko sōdan' *Tokyo Nichinichi Shinbun*, no. 836 (29 October 1874) n.p.
26 Fukuzawa Yukichi, 'Gaikokujin no naichi zakkyo yurusu bekarazaru no ron' in *Minkan zasshi*, 6 (January, 1875), p. 7.
27 Here I mean the cultural influence or value of technological artifacts, not cultural attributes such as the Japanese businessman's so-called samurai ethos and heritage.
28 J. B. Bury, *The Idea of Progress: An Inquiry into its Origin and Growth*, New York: MacMillan, 1932.
29 'The Return of the Embassy,' *The Japan Weekly Mail*, (23 September 1873): 594–5.
30 'Unsound Progress,' *The Japan Weekly Mail*, (30 December 1871): 719–20.
31 Johannes J. Rein, *Industries of Japan*, London: Hodder and Stoughton, 1889, p. 1.
32 Michael Adas, *Machines as the Measure of Men: Science, Technology, and Ideologies of Western Dominance*, Ithaca, NY: Cornell University Press, 1989, pp. 144 and 221.
33 Adas, *Machines as the Measure of Men*, p. 146.
34 Many historians refer to these furnaces as Japanese-style or Ōshima-style blast furnaces.
35 On numerous occasions, simply mentioning my plan to compare the engineers' proposals, the government's technologies, and contemporary practice to Japanese scholars and historians of technology produced reactions such as 'there is nothing wrong with traditional Japanese technology.' Harada Noriko, interview with author, National Museum of Science and Technology, Ueno, Tokyo, July 1996.
36 F.O. 46/125, no. 72. Yokohama General Chamber of Commerce to Sir Harry Parkes.
37 F.O. 46/125, no. 72. Date Home Minister to Parkes, trans., W. G. Aston.
38 The merchant's short-sightedness was so severe that the only thing which helped to curtail the practice was a decline in demand in the mid-1870s. McCallion, *Silk Reeling in Japan*, p. 74; and Ōtsuka, *Sanshi*, 1: 391–2.
39 The French government would probably not have attempted to bully the Japanese to action because at the time, their only interest in Japan was as a matter of prestige, not wanting to be outdone by the British. Historian Richard Sims asserts that Japan was politically and economically too unimportant, and the British trade machine too powerful for the French government to go head to head over an issue such as silkworm egg cards. Richard Sims, *French Policy Towards the Bakufu and Meiji Japan, 1854–95*, Richmond: Japan Library, 1998, p. 21.
40 A representative of the French trading house Hècht, Lilienthal and Co. in 1869 first approached Itō Hirobumi with the idea of a model filature that relied on Western technology.
41 For Japan, Kamaishi was larger and more complex than any previously established Western-style ironworks. Compared to facilities in Europe and North America, Kamaishi was average.
42 The term 'in-blast' denotes a blast furnace that is in operation. To 'blow-in' a blast furnace, first the charge of iron ore and fuel are lit from the bottom and allowed to

slowly kindle. Once they have ignited sufficiently, a small blast of air is introduced via a blowing engine in order to increase the rate of combustion. As the furnace heats up the air volume and pressure – the blast – is slowly increased until full pressure is achieved.

43 As will be discussed in Chapter 5, Matsukata only partly believed in laissez-faire economics. Agro-industries such as animal husbandry, tea production, and sericulture would receive his fullest attention and enthusiastic (economic) support.

44 A partial list of Meiji changes includes development of railway and road systems, a postal system, national mint, national bank, adoption of the Gregorian calendar, patent system, Western-style educational and military systems, telegraph network, system of prefectures, and land and tax reform policies.

45 Many early Meiji intellectuals relied on the published translations of Western ideas. As was also the case with Tokugawa era *rangaku* (Dutch learning) texts, the order in which the books appeared in Japan did not necessarily correspond to the order in which they were published or the ideas were originally presented. For example, Mill's 1859 work *On Liberty* was translated into Japanese in 1871, but Rousseau's 1761 *Social Contract* was not published in Japan until 1882. Kenneth B. Pyle, 'Meiji Conservatism' in *The Cambridge History of Japan*, vol. 5 *The Nineteenth Century*, Cambridge: Cambridge University Press, 1989, p. 676 (full chapter pp. 674–720).

46 Tsuji Zennosuke, *Nihon bunkashi*, vol. 7, Tokyo: Shunjūsha, 1950, p. 18.

47 Donald Shively, 'Nishimura Shigeki: A Confucian View of Modernization,' in Marius B. Jansen, ed., *Changing Japanese Attitudes Toward Modernization*. Rutland, VT: Charles E. Tuttle. 1982, p. 199.

48 See for example: 'Punch to the Volatile Japanese Misguided and Erring Children,' *The Japan Punch*, 1872, reprint edition, vol. 3 (1870–1872) Tokyo: Yushodo, 1975, p. 252, for sarcastic negative assessment of Japanese adoption of Western material culture; 'Japanese Vanity,' *Japan Weekly Mail*, (11 February 1871): 67–9 for comments and call for understanding of Japanese vanity; 'Unsound Progress,' *Japan Weekly Mail*, (30 December 1871): 720–1, for criticism of 'progress' on grounds of superficiality and for abandoning traditional culture.

49 Sugiura Yuzuru and Shibusawa Eiichi, *Kōsei nikki*, xerographically reproduced in Sugiura Yuzuru Iinkai, *Sugiura Yuzuru zenshū*, vol. 5, Tokyo: Sugiura Yuzuru Iinkai, 1979, pp. 195–6 (hereafter *SYZ*); for excerpted translations of Sugiura and Shibusawa's travel diary see Teruko Craig, trans., *The Autobiography of Shibusawa Eiichi: From Peasant to Entrepreneur*, Tokyo: University of Tokyo Press, 1994.

50 *SYZ*, p. 196.

51 Nihon Shiseki Kyōkai, *Ōkubo Toshimichi monjo*, vol. 4, Tokyo: Nihon Shiseki Kyōkai, 1928, p. 468. In this statement, Ōkubo implied that Japan could modernize quickly since it had apparently taken the British a short time to do so.

52 Even witnessing the negative side of industrialization through a tour of some of England's most polluted cities was not enough to deter Meiji modernizers: see Andrew Cobbing, 'Britain [1]: 17 August–16 December 1872, Early Meiji Travel Encounters,' in Ian Nish, ed., *The Iwakura Mission in America and Europe: A New Assessment*, Richmond: Japan Library, 1998, p. 45 (hereafter *Iwakura Mission*).

53 Jacques Maquet has argued that African modernization should not be equated with Westernization any more than European use of paper means that Europe was Sinified by adopting a technology that was originally Chinese: Jacques Maquet, *Civilizations of Black Africa*, New York: Oxford University Press, 1972, p. 170.

54 Shively, 'Nishimura Shigeki,' p. 197.

55 See *OHHS, passim*.

56 Yoshimi Shunya, *Hakurankai no seijigaku: manzanshi no kindai*, Tokyo: Chūōkōronsha, 1992, p. 118.

57 *Nature*, 6 (25 July 1872), p. 250.

58 It is through the records of the *kyōshinkai* that we find numerous discrepancies between private reeling firms' claims to be based on Tomioka's French technology, while descriptions are of Italian or Japanese methods.

59 For a description of Motoda's Confucian-based efforts at protesting *bunmei kaika*, see Donald H. Shively, 'Motoda Eifu: Confucian Lecturer to the Meiji Emperor,' in David S. Nivison and Arthur F. Wright, eds, *Confucianism in Action*, Stanford, CA: Stanford University Press, 1959, pp. 302–33.

60 Itō's reaction was not against the material aspect of *bunmei kaika*. He rejected its liberal ideological component and moved to undermine the Popular Rights Movement. He sought more practical ways of stabilizing the new nation while trying to gain support for the government's policies. See Ivan P. Hall, *Mori Arinori*, Cambridge: Harvard University Press, 1973, pp. 346–47. For a general discussion of the reaction against *bunmei kaika*, see Kenneth B. Pyle, 'Meiji Conservatism,' in *Cambridge History of Japan*, vol. 5, Cambridge: Cambridge University Press, 1989, pp. 674–720.

61 Kenneth B. Pyle, *The New Generation in Meiji Japan: Problems of Cultural Identity, 1885–1895*, Stanford, CA: Stanford University Press, 1969, p. 127. This too, was a time of contradiction within the government. While conservative forces were growing (and becoming outraged), there were still some, like Inoue Kaoru, who continued to publically promote refashioning Japan in to a European-style empire for the sake of treaty revision: see Marius B. Jansen, 'Modernization and Foreign Policy in Meiji Japan,' in Robert E. Ward, ed., *Political Development in Modern Japan*, Princeton, NJ: Princeton University Press, 1968, p. 175. Opened in 1883, the *Rokumeikan*, a Western-style pavilion and dance hall became the conservatives' 'symbol of cultural subservience' Pyle, 'Meiji Conservatism,' p. 689.

62 The government began entertaining ideas of selling Tomioka in 1880, trying to do so unsuccessfully a number of times. After much debate and political wrangling, the government sold the filature in 1893.

63 With the exception of China, Japan, and Korea, by 1895, most of Africa, Asia, the Middle East, and Pacific Islands had fallen under Western domination.

64 Japan's punitive expedition against Taiwan in 1873, its abandonment of Sakhalin and claim to the Kuril Islands in 1874, and 1870s political crisis over Korea, can all be seen as being expressions of national consolidation. Meiji leaders' actions were directed at defining Japan's borders in terms of establishing areas within which it could protect the inhabitants.

65 Itō Hirobumi, Komatsu Midori, ed., *Itō-ko zenshū*, Tokyo: Itō-ko Zenshū Kankokai, 1927, pp. 274–5.

66 Maquet, *Civilizations of Black Africa*, p. 170.

2 Tradition and modernization

1 The following is based on F.O. 46/11/160, Francis Otwell Adams, *Report of a Visit to the Central Silk Districts of Japan*, August, 1869; Stephen McCallion, *Silk Reeling in Meiji Japan: The Limits to Change*,' Ph.D. dissertation, Columbus, OH: The Ohio State University, 1983, pp. 30–5; and Ernest de Bavier, *La Sériciculture le Commerce des Soies et des Graines et L'Industrie de la Soie au Japon*, Lyon: H. Georg, 1874, pp. 128–39.

2 Iida Kenichi, *Nihon tekkō gijutsu*, Tokyo: Tōyō Keizai Shinhō, 1979, p. 33.

3 Melvin C. Aikens and Higuchi Takayasu, *The Prehistory of Japan, Studies in Archaeology*, New York: Academic Press, 1982. p. 255.

4 The first Western visitors to Japan believed that they had come across a country with significant and easily mined mineral resources. This is in stark comparison to the second Western encounter a few centuries later. One Spanish trader writing in 1594

believed that Japan was rich in gold, silver, copper, and iron, that mines were found throughout the country, and that ore extraction was accomplished with ease. Ten years after the Meiji Restoration, however, J. J. Rein, a Professor of Geology at Marburg University, was struck by Japan's apparent lack of mineral resources and previous reports to the contrary.

5 A common fallacy regarding Japan's pre-modern extraction of mineral resources is that no other type of (iron) mining existed other than iron sand. There is a long history of working with iron ore in various parts of the Tōhoku region, especially in the Iwate area. Various relics and artifacts found in this area suggest this, as does a Daido era (806–810) scroll depicting the operation of a *nodatara* furnace.

6 Iida Kenichi, *Tetsu no kataru nihon no rekishi*, Tokyo: Soshiete, 1979, p. 80.

7 Examples of these statements can be found in: *Iida, Nihon no tekkō*; Iida, *Tetsu no kataru*; Iida Kenichi, *Nihon kagaku gijutsushi*, Tokyo: Asahi Shinbunsha, 1964; and Mori Kōichi, *Nihon kodai bunka no tankyū: tetsu*, Tokyo: Shakai Shisōsha, 1974. Nishio Keijiro, 'The Mining Industries of Japan,' *Transactions of the American Institute of Mining Engineers*, 43 (1913), p. 76 states that iron sand was found in the mountainous areas of Izumo, Iwami, Bitchu, Bingo, Bizen and Akio. He also notes that in 750 magnetic iron sand was dug from pits in Mimasaka (p. 58).

8 Iida, *Tetsu no kataru*, p. 81.

9 Johannes J. Rein, *The Industries of Japan*, London: Hodder and Stoughton, 1889, p. 294.

10 Tanimura Hiromu, 'Development of the Japanese Sword,' *Journal of Metals*, 32: 2 (February, 1980), p. 71.

11 Mori, *Nihon kodai bunka*, p. 36.

12 Mori, p. 235.

13 Iida, *Nihon tekkō*, p.23.

14 Iida, *Tetsu no kataru*, p. 84. Quite unlike the Chinese case, where a furnace was typically operated in the winter months because it was not a busy time for farmers, the Japanese chose to operate the *nodatara* during the height of the agricultural season.

15 Iida, *Tetsu no kataru*, p. 83.

16 Iida, *Tetsu no kataru*, p. 83; William Gowland, 'Metals and Metalworking in Old Japan,' *Transactions and Proceedings of the Japan Society of London*, 13 (1915), p. 45.

17 Tanimura, 'Development of the Japanese Sword,' p. 71.

18 Weights are converted from Iida, *Tetsu no kataru*, p. 84.

19 Tanimura, 'Development of the Japanese Sword,' p. 72.

20 Nishio, 'Mining Industries of Japan,' p. 76.

21 For the term and conceptualization of 'foreign knowledge' I rely on Gregory K. Clancey, *Earthquake Nation: The Cultural Politics of Japanese Seismology, 1868–1930*, Berkeley, CA: University of California Press, 2005.

22 *Shokusan kōgyō* is also translated more simply as industrial promotion.

23 The full text can be found in Ōkurashō, ed., *Meiji zenki zaisei keizai shiryō*, vol. 17, Tokyo, 1978. A translation is available in Erich Pauer, ed., *Papers of the History of Industry and Technology in Japan*, vol. 1, Marburg: Förderverein Marburger Japan-Reihe, 1995, pp. lxi–lxii.

24 Sakai cotton mill was a Satsuma-*han* venture, the machinery at Akabane Machine Works was purchased by the Meiji government from Saga-*han*. Although dated Horie Yasuzō, 'Government Industries in the Early Years of the Meiji era,' *Kyoto University Economic Review*, XIV (1939), pp. 67–87, is a good source for basic factual information on early industrialization efforts.

25 *Naimushō* is also translated as Home Ministry and Ministry of the Interior. Tomoko Hashino and Osamu Saito, 'Tradition and Interaction: Research Trends in Modern Japanese Industrial History,' *Australian Economic History Review*, 44:3 (November 2004), p. 244.

26 See for example, Yamazaki Yūkō, 'Nihon kindaika shuhō o meguru sōkoku' in Suzuki Jun, ed., *Kōbushō to sono jidai*, Tokyo: Yamakawa Shuppansha, 2002.
27 On military modernization, see Kozo Yamamura, 'Success Illgotten? The Role of Meiji Militarism in Japan's Technological Progress,' *Journal of Economic History* 37:1 (March 1977), pp. 113–35; also Tessa Morris-Suzuki, *The Technological Transformation of Japan: From the Seventeenth to the Twenty-first Century*, Cambridge: Cambridge University Press, 1994, pp. 79–80.
28 Thomas C. Smith, 'The Introduction of Western Industry to Japan During the Last Years of the Tokugawa Period,' *Harvard Journal of Asiatic Studies*, 11:1/2 (June, 1948), p. 136.
29 Pauer, *History of Industry*, vol. 1, p. lxv. Gooday and Low note that there were over 8,000 foreigners working in Japan, half of whom were Chinese laborers. The Ministry of Public Works alone employed 60 percent of the foreign advisers. See Graeme J. N. Gooday and Morris F. Low, 'Technology Transfer and Cultural Exchange: Western Scientists and Engineers Encounter Late Tokugawa and Meiji Japan,' *Osiris*, 2nd series, 13 (1998–99), pp. 105.
30 Gooday and Low, 'Technology Transfer,' pp. 104–5.
31 Gooday and Low, 'Technology Transfer,' p. 105.
32 The original source on the College of Engineering is: Henry Dyer, *Dai Nippon: The Britain of the East*, London: Blackie, 1904, pp. 89–93.
33 Courses were also taught in German.
34 The best study of Milne, Conder and their work at the Imperial College of Engineering can be found in Clancey, *Earthquake Nation*.
35 Gooday and Low, 'Technology Transfer,' pp. 110–11.
36 Dyer designed a curriculum that combined his own personal experiences and ideas in a format similar to what was found at the Zürich Polytechnic Institute; Gooday and Low, 'Technology Transfer,' p. 109.
37 Richard Henry Brunton, 'Public Works,' *The Japan Mail*, (25 September 1875): 558–564.
38 See for example, Kennichi Ohno, *The Economic Development of Japan: The Path Traveled by Japan as a Developing Country*, Tokyo: GRIPS Development Forum, 2006.
39 Gooday and Low, 'Technology Transfer,' pp. 109–10.
40 Although the Ministry of Public Works is most often credited with the development of Japan's early railroads, it should be noted that initial development was under direct government control. The Ministry of Public Works assumed responsibility after its creation one year later.
41 Dyer, *Dai Nippon*, p. 134.
42 Aoki Eiichi, 'Dawn of Japanese Railways' *Japan Railway and Transport Review* (March, 1994), pp. 28–30.
43 Aoki Eiichi, 'Growth of Independent Technology,' *Japan Railway and Transport Review* (October, 1994), p. 56.
44 Aoki, 'Dawn of Japanese Railways,' p. 29.
45 Aoki Eiichi, 'Expansion of Railway Network,' *Japan Railway and Transport Review* (June, 1994), p. 35.
46 Nakamura Naofumi, 'Meiji-era Industrialization and Provincial Vitality: The Significance of the First Enterprise Boom of the 1880s,' *Social Science Japan Journal*, 3:2 (October 2000) p. 192.
47 Steven J. Ericson, *The Sound of the Whistle: Railroads and the State in Meiji Japan*, Cambridge, MA: Harvard University Press, 1996, p. 99.
48 Aoki,'Dawn of Japanese Railways,' p. 29.
49 Morris-Suzuki, *Technological Transformation*, p. 73.
50 Yakup Bektas, 'Displaying the American Genius: The electromagnetic telegraph in the wide world,' *British Journal for the History of Technology*, 34:121 (June 2001), p. 230.

51 Morris-Suzuki, *Technological Transformation*, p. 73. See Richard Henry Brunton, *Building Japan, 1868–1876*, with introduction and notes by Sir Hugh Cortazzi, Sandgate: Japan Library, Ltd., 1991, pp. 27–29 for Brunton's description of building the telegraph line. Brunton attributes the attacks to samurai needing something to do with their swords.

52 Although there were both public and private railroads until the turn of the century, foreigners were excluded from controlling the railroads.

53 Ericson notes that the Japanese were particularly adept at line construction, having gained technological independence by 1880. It took until the end of the century domestically to manufacture reliable, cost-competitive locomotives. See Steven J. Ericson 'Importing Locomotives in Meiji Japan: International Business and Technology Transfer in the Railroad Industry,' *Osiris*, 2nd series, 13 (1998–99), p. 129.

54 As we shall see with the case of Kamaishi mines and the iron industry in general, Netto and Lyman were not always quite right.

55 Francisque Coignet, *Note sur la richesse minérale du Japon*, in Junkichi Ishikawa, *Nihon kōbutsu shigen ni kansuru oboegaki*, vol. 7, Tokyo: Haneda Shoten, 1944, p. 105.

56 See Curt Netto, *Memoirs of the Science Department, University of Tokio, Japan*. Vol. 2, 'On Mining and Mines in Japan,' Tokyo: University of Tokio, 1879, pp. 7–8.

57 F. G. Notehelfer, 'Japan's First Pollution Incident,' *Journal of Japanese Studies*, 1:2 (Spring, 1975), p. 354. Notehelfer's description of Ashio copper mine in 1877 is essentially the same as what was described nearly a decade earlier.

58 Netto, *Memoirs*, p. 19.

59 Notehelfer, 'Japan's First Pollution Incident,' pp. 355–6.

60 Notehelfer, 'Japan's First Pollution Incident,' pp. 354–5.

61 F. G. Notehelfer, 'Between Tradition and Modernity: Labor and the Ashio Copper Mine,' *Monumenta Nipponica*, 39:1 (Spring, 1984) p. 20.

62 Ōkurashō, Kōbushō enkaku hōkoku, Tokyo: 1889, p. 122, hereafter *KEH*.

63 *Kōbushō enkaku hōkoku, KEH*, pp. 129–31. The government's strategy was to gain the bosses' cooperation by setting good prices for the ore they would purchase. They eventually set a deadline for the end of the system, negotiated a bit more with the bosses so as not to disrupt production, and the system slowly went away.

64 Brian Burke-Gaffney, 'Hashima: The Ghost Island,' *Crossroads*, vol. 4, (Summer, 1996), pp. 34–5.

65 John McMaster, 'The Takashima Mine: British Capital and Japanese Industrialization,' *Business History Review*, 37: 3 (Autumn, 1963), pp. 217–8.

66 McMaster, 'The Takashima Mine,' pp. 224–5. The sale of Takashima mines to Gotō seems to have been a way to appease a man who was mounting an opposition movement against the government following the 1873 *Seikanron*. As one of the founders of the *Aikoku Koto* or Patriot's Party, Gotō began to challenge the Satsuma-Chōshū dominated government by calling for a popularly elected parliament.

67 McMaster, 'The Takashima Mine,' pp. 227–38.

68 Banno Junji, "Fukoku' ron no Seijishiteki kōsatsu,' in Umemura Mataji and Nakamura Takafusa, eds, *Matsukata zaisei to shokusan kōgyō seisaku*, Tokyo: United Nations University Press, 1983, pp. 39–45; Tomoko Hashino and Osamu Saito, 'Tradition and Interaction: Research Trends in Modern Japanese Industrial History,' *Australian Economic History Review*, 44:3 (November 2004), p. 244.

69 Hashino and Saito, 'Tradition and Interaction,' p. 244.

70 Technically the Shinagawa factory was purchased by the Meiji government after it became obvious that entrepreneurs lacked the resources to produce this important 'modern' material. See Martha Chaiklin, 'A Miracle of Industry: The Struggle to Produce Sheet Glass in Modernizing Japan,' in Morris F. Low, ed., *Building Modern Japan: Science, Technology, and Medicine in the Meiji era and Beyond*, New York: Palgrave Macmillan, 2005, pp. 161–81.

71   Kinugawa Taichi, *Honpō Menshi Bōsekishi*, Ōsaka: Nihon Mengyō Kurabu, 1937, pp. 34–6, reprinted in *Meiji hyakunenshi sōsho*, vol. 7, Tokyo: Hara Shobō, 1990; Katō Kōzaburō, 'Yamanobe Takeo and the Modern Cotton Spinning Industry in the Meiji era,' in Erich Pauer, ed., *Papers of the History of Industry and Technology in Japan*, vol. II, Marburg: Förderverein Marburger Japan-Reihe, 1995, p. 5.

72   W. G. Beasley, *Japan Encounters the Barbarian: Japanese Travellers in America and Europe*, New Haven, CT: Yale University Press, 1995, pp. 105–6. Beasley provides a wonderful description of Godai's activities at this time.

73   Janet Hunter, 'Regimes of Technology Transfer in Japan's Cotton Industry 1860s–1890s' paper presented at the ninth conference of the Global Economic History Network, Kaohsiung, Taiwan, 9–11 May 2006, p. 8. I would like to thank Professor Hunter for kindly granting permission to use her paper.

74   Shashi Hensan Iinkai, *Nichibō 75-nen Shi*, Ōsaka: Nichibō Kabushiki Kaisha, 1966, p. 10.

75   Nakamura Naofumi, 'Meiji-era Industrialization,' p. 194.

76   Katō, 'Yamanobe Takeo,' p. 5; Nakamura, 'Meiji Industrialization Boom,' p. 194, n. 7.

77   Gary Saxonhouse, 'A Tale of Japanese Technological Diffusion in the Meiji Period,' *The Journal of Economic History*, 34: 1, p. 152. Saxonhouse goes as far as to assert that Osaka Spinning relied heavily on it British advisors for at least seven years and possibly as long as 35 years.

78   Katō, 'Yamanobe Takeo,' pp. 9–10; Hunter, pp. 21–2.

79   Hunter, 'Regimes of Technology Transfer,' p. 22; Saxonhouse claims that the mill's design was a standard Platt Brothers design and the later selection of ring frames was based more on coincidence than technical evaluation or expertise (p. 153).

80   Katō, 'Yamanobe Takeo,' p. 12.

81   Nakamura, 'Meiji Industrialization,' p. 194; Katō, 'Yamanobe Takeo,' p. 9; Hunter notes that the majority of capital raised by Shibusawa was from the Tokyo area, p. 3.

82   Nakamura, 'Meiji Industrialization,' p. 194; Takeshi Abe, 'Technological and Organizational Absorption in the Development of the Modern Japanese Cotton Industry,' Fifth Conference of the Global Economic History Network, Osaka, Japan, 16–18 December 2004, p.6; Katō, 'Yamanobe Takeo,' p. 12.

83   Serguey Braguinsky, Atsushi Ohyama, and David C. Rose, 'Cooperative Technology Adoption Under Global Competition: The Case of the Japanese Cotton Spinning Industry,' *George J. Stigler Center for the Study of the Economy and the State Working Papers Series*, 27 July 2002, pp. 9–11.

84   Abe, 'Technological and Organizational Absorption,' p. 7; Nakamura, 'Meiji Industrialization,' p. 194.

85   Abe, 'Technological and Organizational Absorption,' pp. 8–17.

86   Hunter, 'Regimes of Technology Transfer,' pp. 3–4.

87   This description of Gaun Tokimune and the *garabō* is based on Morris-Suzuki, *Technological Transformation*, pp. 89–91.

88   Suzuki Jun, 'The Humble Origins of Modern Japan's Machine Industry,' in Masayuki Tanimoto, ed., *The Role of Tradition in Japan's Industrialization: Another Path to Industrialization*, Oxford: Oxford University Press, 2006, p. 145.

89   Unless stated otherwise, information on Toyoda is based on William Mass and Andrew Robertson, 'From Textiles to Automobiles: Mechanical and Organizational Innovation in the Toyoda Enterprises, 1895–1933,' *Business and Economic History*, 25:2 (Winter, 1996) pp. 1–37.

90   Suzuki, 'Humble Origins,' p. 145.

91   *Croisure* or crossing is the process by which silk thread is crossed over itself to help bind and strengthen it. These will be discussed in greater detail in Chapter 3.

92   Tachi Saburō, *Kiito seihō shin'an*. Tokyo, 1874, pp. 13–17. Tachi is also translated as Date and Tate Saburō.

93 Maruyama's boilers were only suitable for low pressure applications.
94 Suzuki, 'Humble Origins,' p. 142.
95 Suzuki, 'Humble Origins,' p. 142.

3 Iron machines and brick buildings

1 Naimushō, *Naimushō dai ikkai nenpō*, Tokyo: 1876, p. 344.
2 Stephen McCallion, *Silk Reeling in Japan: The Limits to Change*, Ph.D. dissertation, Columbus, OH: The Ohio State University, 1983, pp. 183–6. Between 1874 and 1880 at least 69 filatures claimed to be based on Tomioka's technology (p. 262). As will be seen later, any technical similarity during this period was largely superficial.
3 Geisenheimer's given name is not recorded in any writings pertaining to Tomioka. There is also some discrepancy in the record over his country of origin and the company for which he worked. It is generally assumed that he was Dutch and that he was a Dutch commercial official. Although not knowing his true name, Yoshida Mitsukuni states that Geisenheimer was German: Yoshida Mitsukuni *Oyatoi gaikokujin*, vol. 2, *sangyō*, Tokyo: Kajima Kenkyūjō Shuppankai, 1968. His connection with Hècht, Lilienthal and Co. and the commonly used name given to that company, based on its location in the foreign settlement, *Ran-hachi*, have also obscured that connection. *The Chronicle and Directory for China, Japan, and the Philippines for the year 1869*, Hong Kong: Daily Press Office, 1869, p. 79; Yokohama Kaikō Shiryōkan, eds, *Yokohama gaikokujin kyoryūchi*, Yokohama: Yokohama Kaikō Shiryōkan, 1998, p. 91.
4 Ōtsuka Ryōtarō, *Sanshi*, vol. 1, Tokyo: Fusōen, 1900, pp. 245–56; and based on Shibusawa's recollections in *Tomioka seishijō no sōsetsu* reprinted in Tsuchiya Takao, ed., *Shibusawa Eiichi denki shiryō*, vol 2. Tokyo: Shibusawa Eiichi Kankōkai, 1956, pp. 517–8 (hereafter *SEDS*); and Shibusawa Eiichi, *Seien sensei denshokō* reprinted in *SEDS*, p. 522; McCallion, pp. 76–8. See also 'Seien sensei denshi,' reprinted in *SEDS*, p. 522. According to this source, Ōkuma surveyed the officials to see if anyone had any knowledge of the silk industry, and it was he who appointed Shibusawa.
5 Brunat is recorded as working for the company as a clerk beginning in 1867; *The Chronicle and Directory for China, Japan, and the Philippines for the year 1867*, Hong Kong: Daily Press Office, p. 65.
6 See Tomioka Seishijōshi Hensan Iinkai, ed., *Tomioka seishijōshi*, Tomioka: Tomioka-shi Kyōiku Iinkai, 1977, document no. 4, 1:139 (hereafter *TSS*); and Ellen P. Conant, 'The French Connection: Emile Guimet's Mission to Japan, A Cultural Context for *Japonisme*' in Hilary Conroy, Sandra T. W. Davis, and Wayne Patterson, eds, *Japan in Transition: Thought and Action in the Meiji Era, 1868–1912*, London: Associated University Presses, 1984, p. 117; Kamijō Hiroyuki, 'Pooru Buryuna: kikai seishi gijutsu no dokusōteki ishokusha' in *Kōza Nihon gijustu no shakaishi*, Tokyo: Nippon Heironsha, 1986, 2: 12–3, (hereafter *KNGS*).
7 Based on Shibusawa's recollections in *TSS*, also reprinted in *SEDS*, 2: 510–1. *TSS* document no. 4, 1:147–50 for Brunat's proposal; document no. 4, 1:150–2 for his final contract.
8 *The China Directory for 1873*, new series, Hong Kong: China Mail Office, 1873, p. 9. Brunat's name is listed with Hècht, Lilienthal and Co., with the notation of 'Tomyoka' [*sic*] as his location. His period of employment with the Meiji government commenced on 29 November 1870.
9 'Correspondances,' *Moniteur des Soies*, 17 September 1870, n.p.
10 *Ōkuma monjo*, A3995. This is the original version of Brunat's final contract with the Meiji government.
11 Sims, *French Policy*, pp. 220–1.
12 'Correspondences,' *Moniteur des Soies*, 17 September 1870.
13 Sims, *French Policy*, pp. 48–54.

14 F.O. 46/53 *'Confidential enclosure*, Charles Winchester to Earl Russell, 16 March 1865.'
15 F.O. 46/86, *'Confidential memorandum*, Alex von Siebold to Harry Parkes, 11 December 1867'. See also F.O. 46/56 'Confidential No. 29,' John McDonald to Harry Parkes, 19 August 1865, for additional complaints.
16 F.O. 46/56 'Confidential No. 29,' John McDonald to Harry Parkes, 19 August 1865.
17 F.O. 46/56 'Confidential No. 29.'
18 F.O. 46/56 'Confidential No. 29.' In the end, the Comte de Montblanc's efforts were for naught. The resolution was abandoned after he was caught using the spurious title Commissaire General of Japan.
19 F.O. 46/100, 'Alex von Siebold to Edmund Hammond, 28 January 1868'; Sims, *French Policy*, p. 68. Some of the first charges of impropriety against Roches were leveled by a Lyons merchants' association.
20 F.O. 46/56 'Confidential No. 29,' John McDonald to Harry Parkes, 19 August 1865.
21 Sims, *French Policy*, p. 70, n. 90.
22 Louis Gueneau, *Lyon et le Commerce de la Soie*, Lyon: L. Bascou, 1923, pp. 93–4; George J. Sheridan Jr., *The Social and Economic Foundation of Association Among the Silk Weavers of Lyons, 1852–1870*, New York: Arno Press, 1981, 1: 184.
23 McCallion rightly argues that Tomioka's greatest problems stemmed from a lack of vision which often led to conflict and further inefficiency. As a humorous, yet significant example of this lack of foresight, he notes for example that as the facility was ready to open, no one had even considered who would work there! See McCallion, *Silk Reeling in Japan*, pp. 84–92.
24 Because of an earlier adoption into Shibusawa Eiichi's family line, Odaka was technically Shibusawa's first cousin and his brother-in-law; he was Shibusawa's wife, Chiyo's brother, i.e., Shibusawa married his first cousin.
25 Sugiura Yuzuru is also known as Sugiura Aizo.
26 For Brunat's proposal, see *TSS*, document no. 4, 1: 147–50. Steam reeling refers to using steam to heat the cocoon basins, not to the use of steam for generating the power to turn the reeling frames. Reeling is the process by which silk filaments are unwound from the cocoon and made into thread.
27 For the terms of Brunat's final contract, see TSS, document no. 4, 1: 150–2.
28 It should be noted that although the government maintained ultimate authority, Brunat was given considerable responsibility in day-to-day operation of Tomioka. See Brunat's contract in *TSS* and McCallion, *Silk Reeling in Japan*, pp. 79–81.
29 Later Sugiura and Odaka spent a great deal of time and effort to find local sources of coal. Odaka considered finding an alternative to the expensive French coal as one of his more important achievements. See *TSS*, document no. 1, 1: 113–114.
30 Teruko Craig, trans., *The Autobiography of Shibusawa Eiichi: From Peasant to Entrepreneur*, Tokyo: University of Tokyo Press, 1994, pp. 98–108. One of Shibusawa's first moves in office was to recruit men with technical and/or foreign language ability into the bureaucracy. Sugiura was among the recruits.
31 Penelope Francks, *Technology and Agricultural Development in Pre-War Japan*, New Haven, CT: Yale University Press, 1984, p. 11.
32 *TSS*, document no. 233, 1: 539.
33 Odaka blamed Brunat for the choosing the site on which to build the government's model facility. Brunat's reasons were stated to be that the location would not cause hardship to local residents, although some had to be relocated (they were well compensated for their land). Odaka's complaint was not with the remote location, but with the water supply, a key ingredient in silk reeling. Tsukuhara Ryōgo, ed., *Rankō ō*, Tokyo: Takahashi Hiratarō, 1909, pp. 199–202.
34 McCallion, *Silk Reeling in Japan*, p. 82.

35 According to Ōtsuka, the Swiss Consul told Hayami about the problem with Japanese silk. Ōtsuka, *Sanshi*, 1: 256–7.
36 See F. O. 46/111, 'Report upon the Central Silk Districts of Japan,' 7 August 1869, F. O. 46/127, 'Third Report on the Silk Culture in Japan,' 10 August 1870.
37 See F. O. 46/125, 'Yokohama Chamber of Commerce, William Marshall to Sir Harry Parkes,' 11 April 1870.
38 Ōtsuka, *Sanshi*, 1: 257.
39 Odaka Atsutada, 'Seishi no hōkoku,' *Ryūmon zasshi*, vol. 60 (15 May 1893): 5; excerpts of this article are also reprinted in *SEDS*, vol. 2, p. 524. Odaka stated that there were 36 frames, Ōtsuka's description for Maebashi counts 32 reeling frames.
40 McCallion, *Silk Reeling in Japan*, p. 134.
41 Ōtsuka, *Sanshi*, 1: 260–261. This is a partial list of trainees and requests sent to Hayami.
42 McCallion, *Silk Reeling in Japan*, pp. 263–268.
43 One such adaptation was using steam to kill the silkworms, rather than the traditional method of baking them under the sun, which was not always effective. It was necessary to kill the worm before it emerged from the cocoon, by which it would become unreelable; Kannōkyoku, Shomukyoku, *Kyōshinkai hōkoku, kenshi no bu*, Tokyo: Yūrindō, 1880, p. 36 (hereafter *KH*).
44 *KH*, p. 39.
45 *KH*, p. 35, emphasis mine.
46 *KH*, pp. 24–89 *passim*.
47 McCallion, *Silk Reeling in Japan*, p. 267.
48 *KH*, p. 33.
49 Much of the Tomioka mystique was due, in part, to the women who worked there. Some contributed to the myth unknowingly by their criticism of facilities that attempted to replicate Tomioka's ideal, others more directly through their glowing descriptions of the filature itself. In her dairy, Wada Ei described seeing Tomioka as if being in a dream. See Wada Ei, *Tomioka nikki*, Kamijō Hiroyuki, ed., Tokyo: Sōjusha, 1976, pp. 56–8; Kamijō Hiroyuki, *Kinu hitosuji no seishun: 'Tomioka nikki' ni miru Nihon no kindai*, Tokyo: NHK Books, 1978, 21–2.
50 The Akasaka filature is alternatively called the Kankōryō filature, the Kōbushō filature, Tokyo Kankōryō Seishijō (filature), or the Aoi-chō, or Tokyo Aoi-chō filature after its location in Tokyo.
51 Ōkurashō, *Kōbushō enkaku hōkoku*, Tokyo, 1889, pp. 684–8 (hereafter *KEH*).
52 Mueller stayed at the Akasaka Filature until his contract with the Meiji government expired in February 1874. Upon fulfilment of contract, he was formally congratulated for his efforts and given a cash bonus; *KEH*, p. 686.
53 Ōtsuka, *Sanshi*, 1: 317.
54 Ōtsuka, *Sanshi*, 1: 317.
55 *KEH*, p. 685.
56 Sano Tsunetami, *Ōkoku hakurankai hōkokusho: sangyōbu*, Tokyo, 1875, vol. 1, section 6, pp. pp. 423–4, (hereafter, *OHHS*).
57 Hayami Kenzō, 'Hayami Kenzō rireki bassui' in Gunma Kenshi Hensan Iinkai, eds, *Gunma-kenshi*, shiryōhen, vol. 23, Maebashi: Gunma-ken, 1985, p. 273; and Ōtsuka, *Sanshi*, p. 319.
58 Ōtsuka, *Sanshi*, p. 319.
59 Sano Akira, *Dai Nihon sanshi*, Tokyo: Dainihon Sanshi Hensan Jimusho, 1898, p. 422–6.
60 Nakamura Naofumi, 'Kōhatsu koku kōgyōka to chuo-chihō: Meiji Nihon no keiken' in Tokyo Daigaku Shakaikagaku Kenkyūjō ed., *20 seiki shisutemu: 4 kaihatsushugi*, Tokyo: Tokyo Daigaku Shuppankai, 1998, pp. 241–75.
61 *KH*, p. 38.
62 *KH*, p. 86.

63 This, the Yonezawa filature, was supposedly based on copying technologies from Tomioka, Akasaka, and the Nihonmatsu Filature, a Maebashi/Akasaka copy. It used a waterwheel for power and had Italian style machines. The only element that seems to be similar to Tomioka was its size. Originally 50 frames, it was later expanded to 100 frames and employed 200 reelers. *KH*, p. 84.

64 *KH*, p. 31.

65 *KH*, p. 24.

66 *OHHS*, vol. 2, section 1, p. 7.

67 *OHHS*, vol. 2, section 1, p. 7, emphasis mine.

68 Motoyama Yukihiko, trans., George Wilson, 'Meirokusha Thinkers and Early Meiji Enlightenment Thought' in *Proliferating Talent: Essays on Politics, Thought, and Education in the Meiji Era*. Honolulu, HI: University of Hawaii Press, 1997, p. 239.

69 Gregory K. Clancey, *Earthquake Nation: The Cultural Politics of Japanese Seismicity, 1868–1930*, Berkeley, CA: University of California Press, 2006, p. 15. Brick chimneys were also considered representative of the march of industrial progress.

70 This fact was not lost on Shibusawa, who marveled at the steam engines at the International Exhibition in Paris in 1867.

71 Muramatsu Teijirō, 'Basuchan–gunkan to kinu no ito,' *Oyatoi gaikokujin*, vol. 15, *kenchiku–domoku*, Tokyo: Kashimada Shuppankai, 1978, p. 129.

72 Tsukuhara, *Rankō ō*, pp. 204–6

73 Coye does not appear in the *Kōbushō enkaku hōkoku* as a government employee. He is listed, however, as a silk inspector for Mitzu-Bishi Sho-kwai [*sic*] in 1875. See, *The Japan Gazette, Hong Lists and Directory for 1875*, Yokohama: Office of the Japan Gazette, January 1875, p. 51.

74 Nagai Yasuoki, *Seishika hikkei* 3 vols., Tokyo, 1884. The first release of the manuals was in 1878. For details regarding filature construction, see 2: 12 *ff*; for reeling techniques, see 3: 1–31.

75 Nagai, Seishika hikkei, vol. 2, p. 3.

76 Nagai, Seishika hikkei, vol. 2, p. 4.

77 In his argument to replace a wooden bridge in Nagasaki with an iron one, Inoue Kaoru couched his reasoning in terms of strength and permanence: Inoue Kaoru-kō Denki Hensankai, *Segai Inouekō den*, vol. 5. Tokyo: Naigai Shoseki, 1934, pp. 357–8 (hereafter *SID*).

78 Nagai, *Seishika hikkei*, 3: 26.

79 Nagai, Seishika hikkei, vol. 2, p.12.

80 See Giovanni Federico, *An Economic History of the Silk Industry, 1830–1930*, Cambridge University Press, 1997, pp. 106–107; and Yukihiko Kiyokawa, 'Transplantation of the European Factory System and Adaptations in Japan: The Experience of the Tomioka Model Filature' *Hitotsubashi Journal of Economics*, 28 (1987): 29.

81 Nagai, *Seishika hikkei*, 3: 14.

82 Odaka, 'Seishi no hōkoku,' p. 4.

83 Odaka, 'Seishi no hōkoku,' p. 9. 'Shibusawa's proposal' would be the terms of construction elaborated in Brunat's contract which included importing the reeling equipment, steam boilers, and steam engines from France, the government's determination of Tomioka's architecture, and the use of coal for fuel, as well as guidelines for hiring foreign and local workers, their salaries, and terms of employment. See *TSS*, document 4, 1: 150–2.

84 Odaka, 'Seishi no hōkoku', p. 3.

85 Odaka, 'Seishi no hōkoku', p. 8.

86 The 1867 trip to Europe was the third trip for Sugiura. He had been part of Tokugawa missions to Europe in 1861and 1863. During the 1867 mission, he and Shibusawa became close friends. Sugiura's favorable disposition toward Western 'civilization'

may have favorably influenced Shibusawa. Igarashi Akio, *Meiji isshin no shisō*, Kanagawa: Seshoku Shōbō, 1996, pp. 183–7. Also, see Sugiura Yuzuru Iinkai, *Sugiura Yuzuru zenshū*, vol. 1, Tokyo: Sugiura Yuzuru Iinkai, 1979, *passim* (hereafter *SYZ*).

87 Craig, *Autobiography of Shibusawa Eiichi*, p. 94. The *bakufu* asked Shibusawa to accompany the delegation and stay in France because it was felt that he was open-minded and could positively influence Akitake's French education, acting to counterbalance the conservative influences of the guardians and attendants from Mito.

88 Sugiura Yuzuru and Shibusawa Eiichi, *Kōsei nikki*, xerographically reproduced in *SYZ*, vol. 5, pp. 118–119; and *KNGS*, 2: 13.

89 *SYZ*, for Paris and the Exhibition, 5: 119–22 and 5: 194–268, respectively.

90 *SYZ*, vol. 1, pp. 34–5.

91 Igarashi, *Meiji isshin*, p. 183.

92 *SYZ*, 1: 158–9.

93 Igarashi, *Meiji isshin*, pp. 184–5.

94 *SYZ*, 5: 52–78; and Igarashi, *Meiji isshin*, p. 186.

95 Igarashi, *Meiji isshin*, p. 187.

96 It should also be noted that Shibusawa and Sugiura were looking at woven silk textiles in Lyons. There is no mention of raw silk in their journal.

97 For Shibusawa and Sugiura's impressions of Italy, see *SYZ*, 5: 364–88.

98 Sims, *French Policy*, pp. 20–2.

99 *KNGS*, 2: 13.

100 Federico argues that Italy's silk reeling industry led the world in technical innovation and was, by the 1850s, almost completely mechanically self-sufficient; Federico, *Economic History of the Silk Industry*, p. 109.

101 *KNGS*, 2: 13.

102 See Silvana de Maio, 'Italy, 9 May–3 June 1873' in Ian Nish, ed., *The Iwakura Mission in America and Europe: A New Assessment*, Richmond: Japan Library (Curzon Press), 1998, pp. 149–51.

103 de Maio, 'Italy', p. 155.

104 Ōtsuka, *Sanshi*, 1: 320–22.

105 I have found no evidence of similar trips to, or comparisons with, France. Countries such as Italy, Spain, and Belgium (and China) appear to have been considered within the realm of successful competition. France and England raised the bar too high. 'Tomioka seishijo no seihin kekaiteki ni yūmei,' *Tokyo nichi nichi shinbun*, (6 June 1873), n.p.

106 *SYZ*, 5: 373–381. The evaluations of Italy are based only on Shibusawa's recollections; Sugiura had returned to Paris and ultimately Japan the month before.

107 Shibusawa later went on to form a number of companies or other ventures based strictly on his experiences in Paris. These include the Teikoku Genjikō (Imperial Theater), the Tokyo Nichinichi Shinbun, and the Meiji Fire Insurance Company. See Craig, *Autobiography of Shibusawa Eiichi*, pp. 159–62.

108 Umegaki Michio, *After the Restoration*, New York: New York University Press, 1988, p. 15.

109 See Sims, *French Policy*, pp. 31–4, and p. 308, n. 23.

110 Vice Consul Martin Dohmen to F. O. Adams, *Commercial Reports from Her Majesty's Consuls in Japan, 1871*, London: Harrison, 1872, p. 56. Dohmen also erroneously attributes the adoption of Western dress with disarming the samurai: 'The fashions, however, that have hitherto been copied or invented by native tailors, are so varied, and in many instances so ridiculous, as to defy the keenest imagination. This certainly is to be regretted, but the substitution of foreign for native dress has been productive of a most important result, namely the general disarmament of the two-sworded class by peaceful means. The moment the "samurai" began to clothe himself in what he believes to be European garments, he found he could no longer wear his murderous

long swords, and consequently he left them at home' (p. 56). The Japanese adoption
of Western clothing was a constant source of comment and amusement for foreigners,
especially the British, in Yokohama.

111 Inoue to Kido, Summer 1872, in *SID* 1: 520–1.

112 *SID* 1: 523.

113 The text of Hayami's opinion paper is reproduced in Sano Akira, *Dai Nihon sanshi*,
Tokyo: Dai Nihon Sanshi Hensan Jimusho, 1898, p. 323.

114 Sano, *Dai Nihon sanshi*, p. 324.

115 Tsukuhara, *Rankō ō*, p. 6. 'Rickety' was Odaka's choice of words.

116 Tsukuhara, *Rankō ō*, p. 6. By this time Maebashi silk was considered to be high
quality and demanded a high price in international silk markets.

117 Shibusawa Eiichi, 'Seiin sensei denhakkō,' v. 7, section 5, pp. 54–61 (1934–1948),
also reproduced in *SEDS*, vol. 2, p. 522.

118 *OHHS*, vol. 1, section 5, p. 3 and section 6, pp. 7–8 (emphasis mine). In this instance
Sano lumps together silk produced by Tomioka and the Tokyo Kankōryō Filature. At
other times he and the Austrian officials at the exhibition distinguish between the
Tomioka filature and the Tokyo Kankōryō (Akasaka filature) in name only.

119 It is not unlikely that Brunat had in mind the idea of producing silk specifically for
export to the French market at the time he designed Tomioka. He was, after all, a
former employee of a Lyons silk wholesaler. In fact, all the labeling on Tomioka-
produced raw silk was written in French. If production for the French market was
indeed his plan, any facility he proposed needed to manufacture silk of the highest
quality – the quality demanded by the French market. It was only after the expansion
of the American export market years later, which was more forgiving of poor-quality
raw silk, that many local *zaguri* and Western technique-based reelers found a market
for their wares.

120 The text of the directive is reproduced in Ōtsuka, *Sanshi*, 1: 249–250. Merchants and
reelers were rightly accused of corner-cutting and selling inferior quality silk to take
advantage of the newly booming silk export market.

121 Muramatsu Teijirō, *Nihon kindai kenchiku gijutsushi*, Tokyo: Shōkokusha, 1976,
p. 74; Clancey, *Earthquake Nation*, p. 12.

4 Smelting for civilization

1 'Yamao kōbu shōyū kōzan no gi ni tsuku kengen' *Kōbun roku*, 2A–25–678 (4 January
1872), n.p.

2 During the first years of Meiji, the government spent considerable effort promoting
the development of the Kosaka and Ikuno silver mines and the Sado gold mine. See
for example, Saigusa Hirōtō and Iida Ken'ichi, eds, *Nihon kindai seitetsu gijutsu
hattatsushi: Yawata seitetsujo no kakuritsu katei*, Tokyo: Tōyō keizai Shinpōsha,
1957, p. 31; and Curt Netto, *Memoirs of the Science Department, University of
Tokio*, Japan, vol. 2, 'On Mining and Mines in Japan.' Tokyo: University of Tokio,
1879.

3 Inoue Kaoru kō Denki Hensankai, *Segai Inouekō den*, vol. 5, Tokyo: Nagai Shoseki,
1933–34, pp. 357–8.

4 The second bridge, built by Richard Henry Brunton, was also built before Japan
had the necessary economic or industrial resources. Brunton used personal contacts to
import the iron from Hong Kong, and borrowed all the necessary machinery from a
local iron working shop. See Richard Henry Brunton, introduction and notes by Hugh
Cortazzi, *Building Japan, 1868–1876*, Sandgate: Japan Library Ltd., 1991, pp. 35–6,
and 178. Also Richard Henry Brunton, letter in response to Jules Furet, *Japan Weekly
Mail*, (12 March 1870): 189; and Jules Furet, *L'Echo du Japon*, (February 1870).

5 Sano Tsunetami, *Ōkoku hakurankai hōkokusho: kōgyō denpa hōkokusho*, Tokyo:
Ōkoku Hakurankai Jimukyoku, 1875, section 3, pp. 2–3 (hereafter *OHHK*).

6 *OHHK*, section 3, p. 2. Rikuchū-kuni Heigori-gun (present day Iwate prefecture) is the region where Kamaishi village, the ironworks, and mines are located.

7 *OHHK*, section 3, pp.3–4.

8 See for example, Yonekura, Seiichiro, *The Japanese Iron and Steel Industry, 1850–1990: Continuity and Discontinuity*, New York: St. Martin's Press, 1994; or Hanzawa Shūzō, Nihon seitetsu kotohajime: Ōshima Takatō no shōgai, Tokyo: Shinjinbutsu Ōraisha, 1974.

9 Collectively referred to as 'Kamaishi' are five separate blast furnace sites, Ōhashi, Hashino, Kuribayashi, Sahinai, and Sunagowatari, that were developed during the Bakumatsu era.

10 Much of what is written about Kamaishi is derived from two works, Ōshima Shinzō, *Ōshima Takatō kōjitsu*, Hyōgō: Ōshima Shinzō, 1937 (hereafter *OTK*), or Saigusa Hiroto and Iida Ken'ichi, eds, *Nihon kindai seitetsu*. *OTK* is a compilation of Ōshima Takatō's diaries and journals with limited commentary. It also relies on fairly contemporary sources, such as *Kōbushō enkaku hōkoku*, to fill in details around Ōshima's entries. Saigusa and Iida is a pioneering work of sorts. Although not the first book to deal, in part, with Kamaishi Ironworks, the authors attempt to provide an accurate narrative and clear up a number of discrepancies that had appeared since Saigusa began writing about Kamaishi in the early 1940s. (See for example, Saigusa Hiroto, 'Shoki Kamaishi seitetsushi kenkyū oboegaki,' *Kagakushi kenkyū*, 3 (May 1942), pp. 108–19. In the decades since these works have been published, however, new materials have become available that have made it possible to add further detail to the narrative and events surrounding the development of Kamaishi Ironworks.

11 Iida Ken'ichi, *Nihon tekkō gijutsushiron*, Tokyo: Mihari Shobō, 1973, p. 331.

12 According to the account in Hyakunenshi Hensan Iinkai, *Tetsu to tomo ni hyakunen*, vol. 1, Kamaishi City, Iwate: Shin Nihon Seitetsu Kabushikigaisha, 1986, p. 10 (hereafter *TTH*), Ōshima was 19 when he went to Edo to study *rangaku*.

13 The following information on Ōshima's early life is based on: Tamura Eitaro, *Nihon no gijutsusha*, Koa Shobō, 1943, p. 87; Fuji Seitetsu Kabushikigaisha Kamaishi Seitetsujo, *Kamaishi seitetsujo shichiju nenshi*, Kamaishi: Fujiseitetsu Kabushikigaisha Kamaishi Seitetsujo, 1955, pp. 7–8 (hereafter *FSK*); David G. Wittner, *Technology Transfer in the Meiji era: The Case of Kamaishi Iron Works, 1874–1890*, Unpublished Master's Thesis, Columbus, OH: The Ohio State University, pp. 19–20; Yonekura, *Japanese Iron and Steel Industry*, pp. 18–21; and *OTK*.

14 Japanese historians frequently point to the coincidence that Huguenin's text was originally published in the same year that Ōshima was born, 1826. See for example, Mori Kahei and Itabashi Gen, eds, *Kindai tetsu sangyō no seiritsu: Kamaishi seitetsujo zenshi*. Kamaishi: Fuji Seitetsu Kabushikigaisha, 1957. Ōshima translated the work together with Tezuka Ritsuzō; Kobayashi Masaaki, *Nihon no kōgyōka to kangyō haraisage: seifu to kigyō*, Tokyo: Tōyō Keizai Shinpōsha, 1977, p. 57.

15 This event is not mentioned in *OTK*, but is referred to in Ōkurashō, *Kōbushō enkaku hōkoku*, Tokyo: Ōkurashō, 1889 (hereafter *KEH*) and subsequently repeated in company histories such as *FSK*, p. 9 and *TTH*, p. 8. The report goes on to state that Ōshima was authorized to produce iron in Nambu-han from March 1856 before running out of money. This date presents a possible problem, however, as Ōshima was also working in Mito at the same time. Perhaps he was working on these projects concurrently.

16 Ōshima had been introduced to Tokugawa Nariakira, daimyo of Mito-han by Fujita Tōko who was one of Nariakira's advisers. Nariakira had become interested in casting cannon ever since Perry's arrival in 1853, and Fujita provided the connection.

17 Mori and Itabashi, *Kindai tetsu*, p. 68; and *FSK*, p. 8.

18 Ōshima built the furnace with the help of two other samurai, Takeshita Seiemon (Satsuma-han) and Kumada Kamon (Akita-han).
19 Records state that Ōshima produced five 300 monme mortars on the 13th day of the 4th month of Ansei 3 (1856), they were tested 15 days later. *OTK*, p. 9.
20 *FSK*, p. 9.
21 *OTK*, p. 10. According to one source, this was a fierce thunderstorm: *FSK*, p. 8.
22 *OTK*, p. 11.
23 Nambu-*han* shipped 15,000 kanme, or approximately 62 tons (56.25 tonnes) of iron to Edo for the reconstruction project; *FSK*, p. 8.
24 By 1868, there were ten furnaces: Ōhashi (3), Hashino (3), Sahinai (2), Kuribayashi (1), and Sunagowatari (1): see for example, Mori and Itabashi, *Kindai tetsu*, p. 77. According to *FSK*, p. 9, there were three at Ōhashi, two each at Sahinai, Kuribayashi, and Sunagoto, and one at Katsushi. In 1872, two additional furnaces were erected at Ōhashi.
25 J. G. H. Godfrey, born Gottfried Hochstätter (Hochstaetter) on 30 April 1841 in Germany, changed his name to John Gottfried Hochstätter Godfrey after moving to England years later; see Ueda Akikizo, 'Kōzan gichō Gottofurei' in *Showa 60 nendo zenkoku chika shigen kankei gakukyōkai gōdō shūki taikai bunka kenkyūkai shiryō*, G2, Showa 60 [1986] October, pp. 5–6. For Godfrey's contract with the Meiji government, signed 17 October 1871, and renewal contract with the government 9 August 1874; see *Kōbun roku, kōbushō no bu*, 2A: 25: 1219: 4, n.p. Godfrey served as the government's chief geologist and mining engineer from September 1871 until June 1877.
26 Saigusa and Iida, *Nihon kindai seitetsu*, p. 31.
27 See *OTK*, pp. 727–732 for Ōshima's Iwakura Mission diary entries for July 1872.
28 *KEH*, p. 134.
29 Godfrey also offered commentary on copper and silver production.
30 Nihon Kōgyōshi Shiryōshū Hensan Iinkai, eds, 'Watanabe, Wataru, Firudo no-to (1877),' *Nihon kōgyō shiryōshū*, Meiji-hen, 16, mae, no. 1, 1992, pp. 106–7 and 119–20 respectively (hereafter, 'Watanabe'). Watanabe Wataru was a mining engineer for the Meiji government who had studied in Germany and France. Within his field notes for 1877, which survey mining and metallurgical activity and the geology of various parts of Japan, is the only extant copy of Godfrey's notes from his tour during 1872. Godfrey's notes are pages 90–120 of Watanabe's text.
31 'Watanabe', p. 107–8. 'Horse,' also known as 'Bears,' are large ferrous masses that form in the hearths of furnaces that have been in-blast for long periods of time. That the Hashino furnaces were shutting down because of this ferrous accretion in such a short period of time is indicative of other problems, either of a technical nature or related to the quality of fuel, ore, or not using flux. See William H. Greenwood, *A Manual of Metallurgy*, vol. 1, London: William Collins, 1874, p. 112.
32 'Watanabe,' p. 110 and similarly p. 119.
33 'Watanabe', p. 110.
34 'Watanabe', p. 120.
35 'Watanabe', pp. 119–20.
36 Bianchi's contract with the Meiji government was signed on 1 March 1874: 'Oyatoi Doitsujin Biyanshi jōyaku,' *Kōbun roku: Kōbushō no bu*, 2A–25–1216–1, (May 1874); and 'Kōzanryō yatoi Doitsujin kōzanshi seitetsushi Biyanshi to jōyaku o gokansu,' *Dajō ruiten*: gaikoku kōsai, no. 15, 2–72–74 (May 1874).
37 See for example, Saigusa and Iida, *Nihon kindai seitetsu*; Yonekura, *Japanese Iron and Steel Industry*; and Ōhashi Shūji, ed., *Bakumatsu Meiji seitetsuron*, Tokyo: Agune, 1991, p. 279.
38 Saigusa and Iida, *Nihon kindai seitetsu*, for example, state that Kamaishi was based on Yamao's plans and cite the *Kōbushō enkaku hōkoku*. Nowhere in this text, however,

can this claim be substantiated. Ōhashi makes the same claim based on the same supporting evidence: Ōhashi, *Bakumatsu Meiji seitetsuron*, p. 279.

39 For initial proposal see 'Yōkō seitetsu kikai setchu no gi ni tsuku,' *Kōbun roku: Kōbushō no bu*, 2A–25–1214–1 (March, 1874), n.p.

40 'Yōkō seitetsu kikai setchu no gi ni tsuku.'

41 By tramway, Godfrey was describing a rail-based system, not an overhead cable system. His suggestion could have either been for horse-drawn carts or those driven by steam. Because he suggested roads and/or a tramway, the assumption is he was thinking of horse-drawn carts.

42 *KEH*, p. 133.The government decided to appropriate the mines on 12 April 1874, a letter to that effect which also recommended that Ōshima and Koma Rinosuke be appointed managers was written by Itō Hirobumi to Yoshii Tōru on the same day: see *OTK*, p. 771 for a copy of the letter.

43 *FSK*, p. 21.

44 One possible exception was Kanda Kōhei's May 1875 commentary in *Meiroku Zasshi*, 'On Whether Iron Mines Should Be Opened.' Kanda notes the importance of iron import substitution for Japan's nation building efforts and for providing the raw materials for building an iron-clad navy. In his essay he mentions hearing that the government may already be in the process of constructing an ironworks, but this is not confirmed: William R. Braisted, trans., *Meiroku Zasshi: Journal of the Japanese Enlightenment*, Cambridge: Harvard University Press, 1976, pp. 457–60.

45 'Kōko sōdan,' *Tokyo nichinichi shinbun*, no. 1769 (23 Oct. 1877): 982.

46 Frequently, developmental economists try to establish a correlation between Bianchi's proposal and problems with the mill and Ōshima's supposed proposal and the shape of the facility after it was re-opened as Tanaka Ironworks based on the more familiar Ōshima-style blast furnace. For example, see Hayashi Takeshi, *The Japanese Experience in Technology: From Transfer to Self-Reliance*, Tokyo: United Nations University Press, 1990, pp. 95–96.

47 Itō Hirobumi, *Itō-ko zenshū*, vol. 2, Komatsu Midori, ed., Tokyo: Itō-kō Zenshū Kankōkai, 1927, p. 274.

48 Bianchi to Royal Saxon Upper Mining Office for Freiberg, 24 October 1856, OBA 99/80, Bcl. 33, Bl. 179, Bergakademie Freiberg.

49 Verzeichnis der Vorlesungen, Herr Louis Bianchi, OBA 99/04 Bcl. 17, no. 76.

50 Erich Siegfried, *Das Korps Franconia in Freiberg, 1838–1910*, Leipzig: Druck von Breitkopf & Härtel, 1910, pp. 97–8.

51 During the course of his legal troubles, one M. Schültz vouched for Bianchi and another student stating that their 'moral conduct . . . has been beyond reproach': see M. Schültz to Royal Upper Mining Office, OBA 99/04/ Bl. 17, no. 33. Until the matter with the tailor was settled, Bianchi's academic certificates were withheld: see Dr Wolf to Royal Office of Law, District Court, Department of Civil Matters, Freiberg, 29 September 1858, OBA 10799/ Bcl. 3, no. 80 and Dr Wolf to Royal Office of Law, District Court, Department of Civil Matters, 3 January 1859, OBA 10799/ Bcl. 3, no. 85.

52 Godfrey's academic transcripts are xerographically reproduced in Ueda, 'Kōzan gichō Gottofurei,' p. 8.

53 'Oyatoi Doitsujin Biyanshi jōyaku,' *Kōbunroku: Kōbushō no bu*, 2A–25–1216–1, (May 1874); and 'Kōzanryō yatoi Doitsujin kōzanshi seitetsushi Biyanshi to jōyaku o gokansu,' *Dajō ruiten: gaikoku kōsai*, no. 15, 2–72–74 (May 1874).

54 Letter from Yamao Yōzō to Itō Hirobumi, 17 May 1874, xerographically reproduced in Nihon Kōgyōshi Shiryōshū Hensan Iinkai, eds, 'Yamao Yōzō: shokanshū' *Nihon kōgyō shiryōshū*, Meiji-hen, 15, ato, no. 1, Tokyo: Nihon Kōgyōshi Shiryō Kankō Iinkai, 1991, pp. 5–10 (hereafter *Yamao*).

55 Yamao returned to Tokyo from the Heigori region on 6 June 1874: *KEH*, p. 15.

56 Itō, *Itō-kō zenshū*, p. 274.

57 Kobayashi Masaaki, *Yawata seitetsujo*, Tokyo: Hambai Kyoikusha Shuppan Sabisu, 1972, p. 100.
58 *OTK*, p. 783.
59 Ōhashi Shūji claims that the location of the furnaces at Ōhashi, Hashino, and Sahinai was dictated by these locations' proximity to the mines, ample land for laying out the ironworks, and abundant running water: Ōhashi, *Bakumatsu Meiji seitetsuron*, p. 279.
60 And indeed they were: on 24 September 1875, Ōshima's old furnaces at Ōhashi were torn down to make way for the new ironworks. See Saigusa and Iida, *Nihon kindai seitetsu*, p. 35; *OTK*, p. 34.
61 From October 1877 until September 1879 there were more than six serious floods, over a dozen storms that caused some type of damage or construction delay, and at least 16 earthquakes at Kamaishi. There was also frequent fog, rain, heavy snow, and wind storms: Hugh G. Purcell, ed., *Japan Journal: The Private Notes of Gervaise Purcell, 1874–1880*, Alhambra, CA: Meiji Art/Litho, 1975, *passim*.
62 'Watanabe,' p. 119. A *kanme* is equal to approximately 3.75 kilograms or 8.25 pounds.
63 Calculation based on 1000 *kanme* ore and 1600 *kanme* charcoal per furnace to produce 5–600 *kanme* of pig iron per day. Experiments on horse performance in coal mines and as mine surface transportation conducted in 1855 in northern England and Wales demonstrated the use of horses in mining to be economically inefficient. Recognizing variables in wheel diameter, axle type, and rail profile, on a level track, horses were expected to pull between 1 and 4 tons per load, practical ratings dropped appreciably on an incline of 1 foot in 260 (1/8 inch per yard). Effective productivity was also affected by distance. See, Nicholas Wood, 'On the Conveyance of Coals in Underground Coal Mines,' *Transactions of the North England Institute of Mining Engineers*, 3 (1854 and 1855), pp. 239–80; and Frederick Overman, *A Treatise on the Metallurgy*, New York: D. Appleton, 1868, p. 86.
64 Wood, 'Conveyance of Coals,' pp. 239–42.
65 See Yonekura, *Japanese Iron and Steel Industry*, p. 24. In his dissertation, Yonekura translated the same phrase, *oyatoi gaikokujin soncho shugi*, as 'blind obedience' to foreign advisers: Yonekura Seiichiro, *The Japanese Iron and Steel Industry: Continuity and Discontinuity, 1850–1970*, Ph.D. dissertation, Harvard University, 1990, p. 30.
66 Saigusa and Iida, *Nihon kindai seitetsu*, p. 34.
67 Not only was there no such policy, there were policy statements to the opposite effect, and foreign advisers frequently complained about the Japanese government's disregard of their advice: Wittner, *Technology Transfer*, pp. 8–18.
68 Johannes J. Rein, *Industries of Japan*, London: Hodder and Stoughton, 1889, p. 274.
69 'Rikukaigun oyōbi kōbu sanshō gōgi no ue ittai seishisho o sōzō shishashi,' *Kōbun-roku: kōbushō no bu*, 2A–25–1464–3 (4 May 1875). This proposal, which discusses building an ironworks based on Krupp's Essen facility, argues in favor of having Kamaishi serve all of Japan's domestic and military iron producing needs. Along with visits to industrial facilities in England and the United States, Krupp was part of the Iwakura Mission itinerary and was described as 'unparalleled or unrivaled in the world.' See Kume Kunitake, ed., *Tokumei zenken taishi Bei Ō kairan jikki*, Tokyo: Dajōkan Kirokugakari, 1878, 3: 325–330. Also available under the same title in a reprint edition from Iwanami Shoten (1996) with commentary by Tanaka Akira.
70 Michael Adas, *Machines as the Measure of Men: Science, Technology, and Ideologies of Western Dominance*, Ithaca, NY: Cornell University Press, 1989, p. 146.
71 For Yamao's activities in railroad, lighthouse, and telegraph building, see for example: *Kōbun roku, kōbushō no bu*, 2A–25–798 (January through April, 1873). Yamao's

efforts to promote copper mining were the way to pay for 'nation building' through export. See 'Ikuno kōzan kaikō shi,' *Kōbun roku: kōbushō no bu*, 2A–25–798–17 (7 February 1873). n.p.

72 Yoshii Tōru, *Kōgyō yōsetsu*, Tokyo: Yoshii Tōru, 1880. This book was compiled during 1876 and 1877. It is based on Joseph Henry Collins' *First Book of Mineralogy* (1873), *Principles of Metal Mining* (1876), *Principles of Coal Mining* (1876), and William H. Greenwood's *Manual of Metallurgy* (1874).

73 Ōhashi, *Bakumatsu Meiji seitetsuron*, pp. 282–3.

74 See Mori Kahei, *Iwate-ken no rekishi*, Tokyo: Yamagawa Shuppansha, 1973, p. 111; and Ibid., p. 283. It should be pointed out that leather seals were not uncommon and are still used to this day in some applications.

75 *OTK*, p. 33.

76 Interestingly, Yonekura, *Japanese Iron and Steel Industry*, considers this a promotion (p. 24), while Hayashi, *Japanese Experience*, sees it as an 'obvious demotion' (p. 95).

77 The following is from Louis Bianchi, *Kamaishi gichō Bianchi hōkoku*, Kamaishi: 7 June 1875. This document is Bianchi's report on Kamaishi, xerographically reproduced in *TTH*, 2: 413–32 (hereafter, *Bianchi*).

78 Sulphur in iron makes it brittle.

79 David Forbes, 'Report on the Progress of the Iron and Steel Industries in Foreign Countries,' *Journal of the Iron and Steel Institute*, 1875, part. 1, p. 299 (hereafter, Forbes I); and David Forbes, 'Report on the Progress of the Iron and Steel Industries in Foreign Countries,' *Journal of the Iron and Steel Institute*, 1875, part 2, p. 619 (hereafter, Forbes II).

80 The following information related to Purcell comes from a transcription of his dairy, Purcell, *Japan Journal*.

81 'Shipping Intelligence,' *Japan Weekly Mail*, (26 March 1874): 193.

82 Purcell, *Japan Journal*, p. 27.

83 Kuwabara Masa, 'Kamaishi kōzan keikyō,' *Kōgaku sōshi*. no.11 (1882): 539. Kuwabara Masa is alternately known as Kuwabara Sei. He was a government inspector sent to Kamaishi to examine the problems at the ironworks and make recommendations for its repair or disposition. This document is his official report to the government.

84 'Memoirs,' *Minutes and Proceedings of the Institution of Civil Engineers*, XLIX (1877): 273 hereafter *MPICE*.

85 Forbes I, p. 299. Except for brief mention in Wittner, *Technology Transfer*, pp. 49–50, nothing has been written about Forbes and his relationship with Kamaishi. In fact, he has only been known as 'Horubusu' in most texts. Yonekura suggested that this transliteration was possibly 'Forbes' but went no further: Yonekura, *Japanese Iron and Steel Industry*, p. 286, n. 13.

86 *MPICE* (1877), p. 271.

87 William H. Brock, 'David Forbes,' in *New Dictionary of National Biography*, p. 824. I am indebted to Professor Brock for sharing a copy of his pre-published entry with me.

88 *MPICE* (1877), p. 271.

89 *MPICE*, p. 272; Brock, 'David Forbes,' p. 824; John Morris, 'Memoirs of the Late Mr. David Forbes,' *Journal of the Iron and Steel Institute*, 1876, part II, p. 523.

90 Morris, 'Memoirs of the Late Mr. David Forbes,' p. 522; Brock, 'David Forbes,' p. 825

91 Morris, 'Memoirs of the Late Mr. David Forbes,' p. 520.

92 *MPICE*, p. 273.

93 *MPICE*.

94 Forbes I, p. 299.

95 Forbes I, p. 299.

96 For example see, Forbes I, p. 229, and *International Exhibition of 1876: Official Catalogue of the Japanese Section, and Descriptive notes on the Industry and Agriculture of Japan*, Philadelphia, PA: Japanese Commission, 1876, p. 48.

97 Forbes I, p. 299; Forbes II, p. 619, *International Exhibition of 1876*, p. 48.

98 Letter from Yamao Yōzō to Itō Hirobumi (19 November 1877) Itō Hirobumi Kankei Monjo Kenkyūkai, eds, *Itō Hirobumi kankei monjo*, vol. 8, Tokyo: Hana Shobō, 1980, p. 81 (hereafter *IHKM*). Yamao reported in May 1878 that construction was going well and that there were still hopes of finding more charcoal nearby: Letter from Yamao to Itō, (14 May 1878) *IHKM*, p. 82.

99 Forbes II, p. 619.

100 Forbes II, p. 619, emphasis mine: Kuwabara, 'Kamaishi kōzan keikyō,' p. 11: 540.

101 Purcell, *Japan Journal*, p. 28.

102 Purcell, *Japan Journal*, pp. 26–35 *passim*.

103 *Bianchi*, p. 416.

104 Purcell frequently noted the comings and goings of other mine personnel. On occasion he also reported on the bon voyage party festivities.

105 See for example, Saigusa and Iida, *Nihon kindai seitetsu*, p. 35.

106 Itō, *Itō-kō zenshū*, 2: 274.

107 Many of Meiji Japan's foreign employees came to the country to line their pockets. Many had salaries comparable to that of high-level government officials and many made demands on the government that they would never have even thought to have made in their home countries. Bianchi does not fit this category. His salary was meager and his demands all seem to have pertained to improving efficiency at Kamaishi.

108 See for example, Purcell, *Japan Journal*, pp. 30 and 58.

109 Also known as Mori Shigesuke, Mori replaced Koma as manager on 14 September 1875: *KEH*, p. 133: Saigusa and Iida, *Nihon kindai seitetsu*, p. 18. He would leave Kamaishi in 1881 to become Chief Engineer of Japanese Railways and later the President of Japan Railways.

110 Purcell, *Japan Journal*, p. 37.

111 *Bianchi*, p. 424.

112 Casley's initial term of employment with the Meiji government was for two years, beginning on 2 February 1876: 'Kōzanryō e Eijin Kasuree gai ni mei o yatoire,' *Dajō ruiten, 2–72, Gaikoku kōsai*, 15: 46 (19 October 1876); Forbes II, p. 619.

113 William H. B. Casley to Yamao Yōzō, 8 February 1878, original letters are xerographically reproduced in *Yamao*. Also available in *IHKM*, 8: 82.

114 Purcell, *Japan Journal*, p. 37.

115 Purcell stated that there were 10 guests, approximately '30 different dishes, and plenty of Sake [*sic*]': Purcell, *Japan Journal*, p. 37.

116 'Shipping Intelligence–Outwards,' *Japan Times*, (16 February 1878): 142.

117 *Yamao*, Letter from Yamao Yōzō to Itō Hirobumi (14 May 1878) pp. 57–60.

118 'Kamaishi kōzan bunkyokunai yakushitsu no ken' memo from Yamada to Dajōkan sangi, *Kōbun roku: kōbushō no bu*, (22 October 1879) 2A–26–2550–20.

119 'Kamaishi kōzan bunkyoku shiyō shintanzai naegishoku tsuku chiseki no ken,' *Kōbun roku: kōbushō no bu*, (November 1879) 2A–10–2551–8.

120 Kuwabara, 'Kamaishi kōzan keikyō,' 11: 543.

121 Memorandum from Yamao Yōzō to Dajōkan, 'Kamaishi kōzan seitetsu jigyō choshu no gi,' *Kōbun roku: kōbushō no* bu, (29 September 1880) 2A–26–2669–5. According to *KEH*, p. 136, and Saigusa and Iida, *Nihon kindai seitetsu*, p. 54, the furnace was blown in on the 13th. *FSK*, p. 23 states that the furnace was not blown-in until October.

122 *FSK*, p. 23; *KEH*, 135.

123 *FSK*, p. 24.

124 'Yamao Yōzō Sado oyobi Rikuchūkuni Kamaishi kōzan junshi no ken' *Kōbunroku: kōbushō no bu*, (April–June 1881) 2A–26–3080–25.

125 'Yamao Yōzō Kamaishi kōzan bunkyoku no shuppatsu no ken' *Kōbunroku: kōbushō no bu*, (July–August, 1881) 2A–26–3081–13.

126 Memorandum from Yamao Yōzō to Yoshii Tomozane, 'Kamaishi Ōhashi aida tetsudō kisha jimin ken'yō sashiyorusu no ken,' *Kōbunroku: kōbushō no bu*, (13 August, 1881) 2A–26–3081–13.

127 Saigusa and Iida, *Nihon kindai seitetsu*, p. 56.

128 Kuwabara, 'Kamaishi kōzan keikyō,' 11: 540–2.

129 Kuwabara, 'Kamaishi kōzan keikyō,' p. 542.

130 Kuwabara, 'Kamaishi kōzan keikyō,' p. 542.

131 *KEH*, p. 136.

132 James M. Swank, *Statistics of the American and Foreign Iron Trades for 1899*, Philadelphia, PA: The American Iron and Steel Institute, 1900, p. 75; David S. Landes, *The Unbound Prometheus: Technological Change and Industrial Development in Western Europe from 1750 to the Present*, Cambridge, MA: Harvard University Press, 1969, p. 124.

133 Rainer Fremdling, 'Foreign Trade Patterns, Technical Change, Cost and Productivity in the West European Iron Industries, 1820–1870,' in R. A. Church, ed., *The Coal and Iron Industries*, Oxford: Blackwell, p. 346.

134 Based on James M. Swank, *Statistics of the American and Foreign Iron Trades*, Philadelphia, PA: The American Iron and Steel Institute, 1878, p. 82.

135 For over half a century the British failed to erect and operate Western-style ironworks in India. Their inability to successfully exploit Indian iron ore reserves has been blamed on under capitalization, producing pig iron for the export in the face of declining prices, and building ironworks in remote locations. See Daniel H. Buchanan, *The Development of Capitalistic Enterprise in India*, New York: MacMillan, 1934, pp. 274–83; and William A. Johnson, *The Steel Industry of India*, Cambridge, MA: Harvard University Press, 1966, p. 9; and Mahadeva G. Ranade, *Essays on Indian Economics*, Bombay: Thacker & Co., 1898, pp. 158–79.

136 F.O. 46: 191 (22 April and 25 May 1875) 'Mr. Plunkett's Report on the Mines of Japan,' p. 33.

137 F.O. 46: 191 (22 April and 25 May 1875) 'Mr. Plunkett's Report on the Mines of Japan,' p. 33.

138 F.O. 46: 191, (25 May 1875), Letter, Sir Harry Parkes to the Earl of Derby; and F.O. 46: 191, (26 May 1875) Private Memorandum, Sir Harry Parkes to Lord Clarendon.

139 F.O. 46: 191 (22 April and 25 May 1875) 'Mr. Plunkett's Report on the Mines of Japan,' p. 2.

140 Benjamin Smith Lyman, *Geological Survey of Japan: Report on the Second Year's Progress of the Survey of the Oil Lands of Japan*, Tokyo: Public Works Department, 1878, p. 47. Both the Catalan and bloomery furnaces are primitive devices, the former developed in the thirteenth century.

141 Lyman, *Geological Survey of Japan*, pp. 47–8. Lyman based his estimate of 5,000 tons per year on a Ministry of Public Works estimate of requiring approximately 10,000 tons of ore per year for the furnaces. He also based his calculations on private capital at 18 percent interest, which he noted should be lower with government funds.

142 Kuwabara, 'Kamaishi kōzan keikyō,' 11: 543.

143 F.O. 46: 191 (22 April and 25 May 1875) 'Mr. Plunkett's Report on the Mines of Japan,' p. 24.

144 Henry S. Monroe, 'The Mineral Wealth of Japan,' *Transactions of the American Institute of Mining Engineers*, 5 (1876–77) p. 269.

145 'Export Trade,' *North China Herald*, 5 February 1874, p. 123; and 'Export Trade,' *North China Herald*, 22 August 1874, p. 205.

146 Kuwabara Masa, 'Kamaishi kōzan keikyō hōkoku,' *Kōgaku sōshi*, no. 10 (1882), p. 488.

147 J. G. H. Godfrey, 'Notes on the Geology of Japan,' *The Quarterly Journal of the Geological Society of London*, 34 (1878): 532–533.

148 Lyman, *Geological Survey of Japan*, p. 46.

149 *Bianchi*, p. 426.

150 Rinji Seitetsu Jigyō Chōsa Iinkai, *Kamaishi oyobi Sennin tetsuzan junshi hōkoku*, Tokyo, 1893, pp. 9–19 (hereafter *RSJCI*). After Kamaishi's sale and resurrection as Kamaishi Mines, Tanaka Ironworks, the Meiji government sent doctor of mining and metallurgy Noro Kageyoshi to investigate the causes of Kamaishi's failure. In his report, Noro vindicated Godfrey by stating that the problem was not related to a lack of natural resources, as had been previously asserted. Rather, he argued that inexperience was the greatest problem.

151 Netto, *Memoirs of the Science Department*, p. 37.

152 Netto, *Memoirs*, pp. 37–40.

153 In addition to Kamaishi Ironworks, the government purchased another privately operated facility at Nakakosaka, in present day Gunma prefecture, in 1878.

154 'Rikukaigun oyōbi kōbu sanshō gōgi no ue ittai seishisho o sōzō shishashi,' *Kōbunroku: kōbushō no bu*, 2A–25–1464–3 (4 May 1875).

155 Swank, *American and Foreign Iron Trades* (1878), pp. 83–7.

156 'Unsound Progress,' *The Japan Weekly Mail*, (30 December 1871): 719–21.

157 Richard H. Brunton, 'Public Works,' *The Japan Mail*, (25 September 1875): 558–64.

158 'Punch to the Volatile Japanese Misguided and Erring Children,' *The Japan Punch*, (reprint edition), p. 252.

159 'Mining Productions,' *Chūgai Bukka Shinpō*, reprinted in *The Japan Weekly Mail*, (23 November 1878): 1271.

160 An 11 September 1880 issue of *The Japan Weekly Mail* noted that Hasegawa Yoshimichi had left on an inspection tour of Anin, Innai, and Kamaishi mines (p. 1180). There are two other articles which mention Kamaishi Ironworks: 'Kamaishi,' *The Japan Weekly Mail*, (28 August 1875): 740–42, which is about sightseeing in the area; and another in the 5 April 1876 issue of *Naniwa shinbun*, which briefly discusses Kamaishi's railroad. There is no coverage in the *Tokyo nichinichi shinbun* or the *Yūbin hōchi shinbun* from 1874 through 1880.

161 J. Arthur Phillips, *Elements of Metallurgy, A Practical Treatise on the Art of Extracting Metals from the Ores*, London: Charles Griffin, 1887, p. 216.

162 One of the hallmarks of 'civilization' was precision. In this sense Kamaishi's steam engines and other mechanical appliances fit the definition. Without an operational furnace, however, these exemplars of 'civilization' were useless and would only be reminders of failure.

163 The full text of Itō's report is reproduced in Saigusa and Iida, pp. 65–6. The total cost of Kamaishi was ¥2,376,625. See Ministry of International Trade and Industry, Tsūshō Sangyōsho, ed., *Shōkō seisakushi: gyōsei kiko*, vol. 17, Tokyo: Shōkō Seisakushi Kankōkai, 1962, p. 19; Kobayashi, *Nihon no kōgyōka to kangyō haraisage*, p. 138.

164 See Obana Fuyukichi, 'Seitetsujo kensetsuron,' reproduced in *Itō Hirobumi hisho ruisan: jitsugyō, kōgyō shiryō*, Tokyo: Hisho Ruisan Iinkai, 1936, p. 75; and Ōkurashō, *Tekkō*, Tokyo: 1892.

165 Kamaishi's railroad's track, locomotives and some to the accessories were removed from the site and sent to Sakai City, Osaka, for use there. The land was returned to the public domain under Iwate prefecture's jurisdiction. Contrary to the wishes of the Ministry of Home Affairs, the machinery was given over to three individuals and all remaining fuel was sold. See *KEH*, p. 137; and Kobayashi, *Nihon no kōgyōka to kangyō haraisage: seifu to kigyō*, p. 211. Steven Ericson describes Japan's measures to 'economize' railroad construction citing contemporary British observations that

Japanese rails were pieced together from scraps of metal. See Steven Ericson, *Sound of the Whistle: Railroads and the State in Meiji Japan*, Cambridge, MA: Harvard University Press, 1996, p. 80 and n. 215.

166 *Segai Inoue-kō den*, 1: 520–1.
167 Ericson, *Sound of the Whistle*, p. 104.
168 Saigusa and Iida, *Nihon kindai seitetsu*, p. 30.
169 Japan relied on imports of British rails through the 1880s, largely switched to the cheaper German and American stock in the 1890s and continued to rely on foreign rails until the 1920s despite opening the Yawata Ironworks in 1901. The first Japanese locomotive was built in 1893 and until that time, the country relied solely on imported engines from Britain and later the United States. See Ericson, *Sound of the Whistle*, p. 32.
170 Itō, *Itō-kō zenshū*, 2: 273–4.
171 Itō, *Itō-kō zenshū*, 2: 274–5. Itō refers to making no progress in the iron industry, and not being able to expand Japan's stores of large weapons, iron warships, or railroads. For Itō, the turning point was the Sino-Japanese War.
172 See n. 140 above.
173 Ōsaka Hōhei Kōshō, 'Itaria 'Guregorini-'chūtetsu narabi ni honkoku Kamaishi chūtetsu o motte seizōseru dangan no hikakushiki hōkoku,' in Ōkurashō, *Tekkō*, pp. 149–52. Also reproduced in Meiji Bunkō Shiryō Iinkai, ed., *Meiji zenki sangyō hatatsushi shiryō*, 70: 4, Tokyo: Meiji Bunkō Shiryō Iinkai, 1971. Recent studies have shown that Tanaka's iron was probably not of equal quality to the material produced at the Italian factory. Takamatsu Tōru has examined old furnace apparatus and argues that defects in the machinery would have made production of high-quality iron doubtful; see Takamatsu Tōru, 'Kamaishi Tanaka seitetsujo mokitankoro no tekkan neppurō,' *Gijutsu to Bunmei*, 10 (June, 1990), pp. 47–67.
174 *RSJCI*, pp. 15–19. Noro's views are also printed in Noro Kageyoshi, 'Honkoku seitetsugyō no kako oyobi shōrai,' *Tetsu to hagane*, 1(1916), pp. 1–16.
175 Rein, *Industries of Japan*, p. 292.
176 Richard Henry Brunton, 'Public Works,' *The Japan Mail*, (25 September 1875): 558–64.
177 Based on his observations at the 1873 International Exhibition in Vienna, Austria, Sano Tsunetami classified the nations of Europe by level of industrialization as displayed at the exhibition. He ranked England and France first, followed by Germany, Austria, Holland and Belgium. See *OHHS* 2: 1–2.
178 Ishii Kendo, *Meiji jinbutsu kigen*, vol. 6, Tokyo: Sakuma Gakugei Bunko (1908) 1997, p. 148. Shibusawa purchased equipment for drying and testing raw silk that sat in an empty warehouse for a number of years until the Office for the Promotion of Industry took it and set it up in the Yotsuya Experimental Station in 1875.
179 The author argued that allowing Japanese officials to wear traditional ceremonial clothes would decrease dependence on the importation of expensive Western clothing. *Hakama* are traditional long pants often described as a divided skirt. 'Japanese Dress for the Japanese,' *The Japan Times*, (6 April, 1878): 292, reprinted from *Hōchi Shinbun*. The government's move to abandon the use of leather boots for the army was similarly couched in terms of economics. Stating that the use had 'originated in the military systems of Europe,' the government intended to give up the practice as part of its effort to reduce the importation of foreign goods. The author of this article went on to rationalize the government's position, stating that if the whole issue was based on appearance, that one would similarly have to give up cotton and woolen cloth uniforms for ones made of the finest woolen fabrics – 'there would be no end to the expense.' 'On Abandoning the use of Leathern Boots for Japanese Soldiers,' *The Japan Times*, (2 February, 1878): 95, reprinted from *Hōchi Shinbun*.
180 This figure is low because there are no records for pig iron production from iron sand in traditional furnaces. It should also be noted that this total amount is considered

low-level production. Ministry of International Trade and Industry, *Shōkō seisakushi: tekkōgyō*, 17, pp. 8–11; and Netto, *Memoirs of the Science Department*, pp. 32–40.

181 After nearly a decade of declining iron prices, British pig iron could be purchased exclusive of transportation charges for as little as ¥11.6 to ¥13 per ton. 'Export Trade,' *North China Herald*, (25 May 1880): 467. The cost of pig iron produced at Kamaishi was ¥35 per ton; see n. 142 above. Bar iron was much closer in price. The Japanese product sold for ¥7.9 per picul, or approximately ¥118 per ton, while foreign bar iron fetched ¥6.24 per picul or approximately ¥98 per ton. See Gaimushō, ed., 'Remarks on Draft Tariff Protocol no. 6, 26 March 1882, *Nihon gaikō bunsho*, vol. 2, Tokyo: Gaimushō, 1964, p. 69.

182 Matsukata Masayoshi, *Zaisei kanki gairyaku*, Tokyo: 1880, reprinted in *Meiji zenki zaisei keizai shiryō shūsei*, vol. 1, Tokyo: Meiji Bunken Shiryō Kankōkai, 1962, p. 534. (Hereafter *MZZKSS*).

183 James Bartholomew notes that Matsukata 'came to share Ōkubo's interest in animal breeding techniques.' Through the pair's control of the Ministries of Finance and Home Affairs, a great deal of money was channeled to support their projects. See James R. Bartholomew, *The Formation of Science in Japan*, New Haven, CT: Yale University Press, 1989, p. 128 and p. 315, n. 23.

184 See Tokutomi Sohō, *Kōshaku Matsukata Masayoshi den*, vol. 1, Tokyo: Kōshaku Matsukata Masayoshi Denki Hakkōjo, 1935, pp. 711–32; and Maeda Masana, *Kōgyō Iken*, in *MZZKSS*, vol. 18 p. 436.

185 After a series of confrontations with China over Korean autonomy in the mid- and late-1880s, the issue of building an ironworks, this time for steel production, was revisited. Although initially popular, political turmoil in the Diet prevented the passing of a bill to fund an ironworks in 1892. Efforts late in that and the following year led by Itō Hirobumi and subsequently Goto Shōjirō were similarly defeated. It was not until China's defeat after the first Sino-Japanese War in 1895 that the government finally approved measures to fund a state-owned iron and steel works.

5 Bunmei kaika to gijutsu

1 Matsumura Harusuke, *Kaika senjimon*, Tokyo: Bunkeidō, 1873. A literal translation of this title would be 'one thousand words of civilization.' Alternate pronunciation for author's name (Harusuke) is Shunsuke.

2 Matsumura, *Kaika senjimon*, 'Shogen,' n.p.

3 That is, *tetsudō, kōzan, kannō, seishi, rengaishi, denshin, doboku, shōki*, and *kikai*: Matsumura, *Kaika senjimon*, pp. 2–4. Matsumura also included the names of the new ministries, such as the Ministry of Public Works (*Kōbushō*) and of the military branches such as the navy (*kaigun*).

4 Matsumura, *Kaika senjimon*, pp. 6–7. He does not seem to have been influenced by Nakamura's translation of Mill's *On Liberty* (1870). For the influence of *Self Help* on Japan see: Earl H. Kinmonth, *The Self-Made Man in Meiji Japanese Thought: From Samurai to Salary Man*, Berkeley, CA: University of California Press, 1981.

5 Matsumura, *Kaika senjimon*, pp. 10 and 12. It is obvious that prostitution and bribery were not new to Japan. I find it interesting, however, that Matsumura chose to include these words in a reader dedicated to teaching the vocabulary used for describing artifacts, institutions, and ideas which were largely new to Japan.

6 I use the word absorb (absorption) instead of either adopt or adapt because it is neutral with regard to technical or cultural modification.

7 For criticism that Japan's 'progress' was superficial see, for example: 'Unsound Progress,' *Japan Weekly Mail*, (30 December 1871): 720–1.

8 In his *Bunmeiron no gairyaku*, Fukuzawa makes frequent reference to stages of 'civilization' and the ability to advance through the stages of historical development as a nation moves toward the attainment of 'civilization.' See, for example, David A.

Dilworth, and G. Cameron Hurst, trans., *Fukuzawa Yukichi's An Outline of a Theory of Civilization*, Tokyo: Sophia University Press, 1973, p. 171.

9 Alexander von Siebold, *Japan's Accession to the Comity of Nations*, London: Kegan Paul, Trench, Trübner & Co. Ltd., 1901, p. 15. von Siebold discussed Japan's ability to enter into diplomatic relations with some degree of parity with the West through a backdoor approach of trade related treaties such as postal unions or telegraph conventions.

10 Schatzberg argues that 'technologies, are, . . . cultural expressions, reifications of human purposes,' and that 'the symbolic meanings of technical things do more than shape modern culture; they also influence the course of technical change itself.' For Meiji cultural materialists, technological artifacts were more than simple representations of 'progress,' they *were* 'progress.' Choice of technique, and even the technology one chose to advertise, were determined by the symbolic meaning of the machine or its material. See Eric Schatzberg, *Wings of Wood, Wings of Metal: Culture and Technical Change in American Airplane Materials, 1914–1945*, Princeton, NJ: Princeton University Press, 1999, p. 232.

11 As early as 1871, the British were doubting the Japanese ability to comprehend the 'spirit' of Western liberalism. One author argued that the Iwakura Mission would be a waste of time because the officials would not truly understand the West: 'The Grand Tour,' *Japan Weekly Mail*, (15 July 1871): 389–91. Similar criticism continued throughout the first decade of Meiji. See for example, 'Modern Japan,' *Japan Weekly Mail*, (28 May 1878).

12 'Kōko sōdan,' *Tokyo nichinichi shinbun*, no. 836 (29 October 1874) n.p.

13 Tomita Masafumi and Tsuchibayashi Shunichi, eds, *Fukuzawa Yukichi zenshū*, vol. 4, Tokyo: Iwanami Shoten, 1959, p. 373, (hereafter *FYZ*); Fukuzawa, *Minkan keizairoku*, vol. 2 (1880); Sugiyama Chūhei and Hiroshi Mizuta, *Enlightenment and Beyond: Political Economy Comes to Japan*, Tokyo: University of Tokyo Press, 1988, p. 46.

14 Matsuzawa Hiroaki cites examples of 'beef, bowler hats, gas lamps, the telegraph, railways, and steamship' as some of the Western components that exemplified *bunmei kaika*. Matsuzawa Hiroaki, 'Varieties of *Bunmei Ron*' in Conroy, Hilary, Sandra T. W. Davis, and Wayne Patterson, eds, *Japan in Transition: Thought and Action in the Meiji Era, 1868–1912*, Rutherford: Fairleigh Dickinson University Press, 1984, p. 209. Kosaka Masaaki describes the extremes to which people would go to equate anything Western with progress and civilization. He quotes a popular satirical song written after an 1871 edict on abandoning the traditional Japanese hairstyle:

> Rap a shaven head and out comes the cry:
>   'Vacillating! Irresolute!'
> Pound on a busy-haired head and it cries:
>   'Restore Imperial Rule!'
> But tap a Western-style-coiffured head and you hear:
>   'Civilization! Enlightenment!'

Kosaka Masaaki, ed., David Abosch, trans., *Japanese Thought in the Meiji Era*, Tokyo: Pan-Pacific Press, 1958, pp. 57–8.

15 Fukuzawa Yukichi, *Bunmeiron no Gairyaku, in FYZ*, 4: 202–4; Dilworth, pp. 13–14.

16 Dilworth, p. 14.

17 Jiro Kumagai, 'Enlightenment and Economic Thought in Meiji Japan: Yukichi Fukuzawa and Ukichi Taguchi,' in Shiro Sugihara and Toshihiro Tanaka, eds, *Economic Thought and Modernization in Japan*, Northampton, MA: Edward Elgar, 1998, pp. 26–7.

18 Michael Adas, *Machines as a Measure of Men, Science, Technology, and Ideologies of Western Dominance*. Ithaca, NY: Cornell University Press, 1989, pp. 133–53, and 221.

19 For Ōkubo's views on Japanese abilities to industrialize rapidly see: Ōkubo Toshimichi to Ōyama Iwao, 20 December 1872, *Ōkubo Toshimichi monjo*, 4: 467–70. During his visit to England, Europe, and the United States, Ōkubo concluded that the source of England's wealth and power was its industry; and that the greatest extent of its industrial growth took place in only the last 50 years. For Kume's comments on the recency of Western achievements, see Kume Kunitake, *Tokumei zenken taishi Bei Ō kairan jikki*, vol. 3, Tokyo: 1878, pp. 325–30; Marlene Mayo, 'The Western Education of Kume Kunitake, 1871–6,' *Monumenta Nipponica*, 28: 1 (1973), pp. 3–67; and Eugene Soviak, 'On the Nature of Western Progress: The Journal of the Iwakura Embassy' in Donald H. Shively, ed., *Tradition and Modernization in Japanese Culture*, Princeton, NJ: Princeton University Press, 1971.

20 Fukuzawa Yukichi, *Minjō isshin*, (1879) in *FYZ*, 5: 8.

21 Kōsaka Masaaki also identifies a distinctive first decade of the Meiji era, 1867–77, which he calls the *Meirokusha* phase. He argues that this was a time when Japan 'adopted Western civilization *in toto*; their purpose was to educate and enlighten the Japanese.' Kōsaka, *Japanese Thought in the Meiji Era*, p. 134.

22 In the late 1880s, Tokutomi Sohō described how earlier domestic reforms had been carried out in the name of treaty revision, that is foreign policy. Although considering them beneficial for Japan, he noted that the needs of '7,000 foreign residents had outweighed the influence of the 38 million Japanese people in determining the recent course of the nation.' See Kenneth B. Pyle, *The New Generation in Meiji Japan: Problems of Cultural Identity, 1885–1895*, Stanford, CA: Stanford University Press, 1969, p. 105.

23 Gregory K. Clancey, *Earthquake Nation: The Cultural Politics of Japanese Seismicity, 1868–1930*, Berkeley, CA: University of California Press, 2006, p. 97. Many of the ministerial buildings were not completed until after the first period of 'Westernization.' The seeds, however, were planted at the earliest stages. Clancey notes how British architect Josiah Conder wrestled with formulating a style of new Japanese architecture. As the government's architect, he was required to design 'accurate Western-style ministerial buildings and country houses' for the government as an expression of their new authority and status. As an architect, however, Conder searched for a Japanese aesthetic to ground his designs in Japanese tradition. In the end, he settled for an 'Indian–Chinese–Japanese' style largely reflective of British colonial designs found in India and elsewhere in the empire.

24 Nagai Yasuoki, *Seishika hikkei*, vol. 2, Tokyo, 1884, p. 4.

25 Hosokawa Junjirō, *Seishi hitsukei*, Nihonbashi: Yamashiroya, ca. 1875, p. 4. Hosokawa continues: 'if the architecture is grand and regulations followed, the quality of the silk will be good and [the filature] profitable.' On Hosokawa, see: Asahi Shinbunsha, eds., *Nihon rekishi jinbutsu jiten*, Tokyo: Asahi Shinbunsha, 1994, pp. 1520–1.

26 Steven Ericson, *Sound of the Whistle: Railroads and the State in Meiji Japan*, Cambridge, MA: Harvard University Press, 1996, p. 98. Harada Katsumasa states that 'the government authorities aimed to shock Japanese citizens with an element of Western civilization and strengthened the centralized administration by making use of the railways.' See Harada Katsumasa, 'The Technical Progress of Railways in Japan in Relation to the Policies of the Japanese Government,' in Erich Pauer, ed., *Papers of the History of Industry and Technology in Japan*, vol. 2, Marburg: Förderverein Marburger Japan-Reihe, 1995, p. 185.

27 Ericson, *Sound of the Whistle*, p. 98.

28 Ōtsuka Takematsu, ed., *Iwakura Tomomi kankei monjo*, vol. 8, Tokyo: Nihon Shiseki Kyōkai, 1935, pp. 114–15. Saigō Takamori was one of the military officials who recommended *against* government expenditure on railroad development. He argued that the government should dedicate its limited financial resources toward building Japan's military; railroads and other public works projects were considered secondary.

Only after the Seinan War of 1877 would Meiji officials fully realize the strategic importance of railroads for the military. Also see Ericson, *Sound of the Whistle*, pp. 98–9.

29 Ericson, *Sound of the Whistle*, pp. 99–100.
30 Kōsaka, *Japanese Thought in the Meiji Era*, p. 56.
31 Charles Wentworth Dilke, 'English Influence in Japan,' *Fortnightly Review*, vol. 20, new series, (1 July to 1 December, 1876): 441.
32 *Yūbin hōchi shinbun* (April 1878): n. p.
33 Murakami Shigeyoshi, ed., *Seibun kundoku kindai shōchokushū*, Tokyo: Shin Jinbutsu Ōraisha, 1983, pp. 70–1; Kunaicho, eds., *Meiji tennōki*, vol. 2, Tokyo: Yoshikawa Kōbunkan 1969, pp. 531–2. One year later, on 12 November 1872, the government issued a statement which ordered Western-style dress for all court and official ceremonies. See Yanagida Kunio, ed., *Japanese Manners and Customs in the Meiji Era*, Charles S. Terry, trans., Tokyo: Pan-Pacific Press, 1957, p. 11.
34 'Japanese Dress for the Japanese,' *The Japan Times*, (6 April, 1878): 292. In this article, the author criticizes the government for its adoption of Western-style clothing.
35 By the end of the first period of Meiji Westernization, the government moved to abandon the use of leather boots and woolen uniforms. See, 'On Abandoning the use of Leathern Boots for Japanese Soldiers,' *The Japan Times*, (2 February, 1878): 95.
36 Dilke, 'English Influence in Japan,' p. 435.
37 As with many of the early popular displays of Westernization, this point was an especially easy target for Meiji satirists. Popular author Kanagaki Robun, for example, frequently lampooned those who flocked to Western fashion and food. In his popular 1871 social satire, *Aguranbe*, Kanagaki chided: 'Samurai, peasants, artisans, merchants, young and old, men, women, wise men, fools, rich, poor, all of these, so long as they eat meat are not culturally backward.' Based on translation in Kōsaka, *Japanese Thought in the Meiji Era*, p. 57.
38 'Kōko sōdan,' *Tokyo nichinichi shinbun*, no. 603 (8 February 1874) n. p.
39 'Kōko sōdan,' *Tokyo nichinichi shinbun*.
40 Sano Tsunetami, *Ōkoku hakurankai hōkokusho: sangyōbu*, vol. 2, Tokyo: Ōkoku Hakurankai Jimukyoku, 1875, section 1, p. 6 (hereafter *OHHS*).
41 *OHHS*, 1: 6.
42 *OHHS*, 1: 7.
43 Yoshimi Shunya, *Hakurankai no seijigaku: manzanshi no kindai*, Tokyo: Chūōkōronsha, 1992, pp. 115–16.
44 Hugh Borton, *Japan's Modern Century*, New York: The Ronald Press, 1955, p. 113.
45 The Meiji government relied on the help of Kurt Wagner, a German national, to select the items that would best display Japanese culture to the world. In addition to silk from Japan's revamped (revitalizing?) reeling industry, Wagner selected other items such as fine porcelains, woven goods, and a model of the Kamakura Buddha. See Yoshimi, *Hakurankai no seijigaku*, pp. 116–17.
46 Yoshimi, *Hakurankai no seijigaku*, pp. 117–21.
47 Peter F. Kornicki, 'Public Display and Changing Values: Early Meiji Exhibitions and Their Precursors,' *Monumenta Nipponica*, 49:2 (Summer, 1994), p. 190. Kornicki notes that post-Restoration exhibitions were official events, unlike their Tokugawa-era forerunners. Meiji-era exhibits, however, were symbolic of Japan's new internationalism and were imbedded within the larger system which 'equated westernization with 'civilization'' (p.194).
48 *Nature*, 6 (25 July 1872): p. 250.
49 'Hōkoku: Kōbushō go-kōnai seishijo,' *Tokyo nichinichi shinbun*, no. 813 (2 October 1874) n.p.
50 'Hōkoku: Kōbushō go-kōnai seishijo.'
51 As argued in Chapter 3, many reelers claimed to have adopted Tomioka's methods in an attempt to demonstrate the status of their filatures and superiority of the silk and

methods. Most, however, had retained largely traditional reeling methods. Gunma prefecture was one of the last to adopt Western reeling methods despite the presence of Tomioka Silk Filature.

52 Richard Henry Brunton tried a number of lighthouse designs to ensure the ability of the lights to remain intact and on course after an earthquake. Some of these included storied structures where the upper level containing the light was isolated from the foundation with rollers. None of Brunton's designs proved effective and he became a proponent of the school of architectural and engineering design that (erroneously) believed that massive, solid buildings were better at resisting an earthquake's destructive forces. See Richard Henry Brunton, 'The Japan Lights,' *Proceedings of the Institution of Civil Engineers*, vol. xlvii, 1876–77. Also reproduced in Brunton, *Building Japan*, Sandgate: Japan Library, Ltd., 1991, pp. 233–67.

53 *OHHK*, p. 2.

54 See government directive in Ōkurashō, *Kōbushō enkaku hōkoku*, Tokyo: Ōkurashō, 1889, p. 133 (hereafter *KEH*).

55 Tsūshō Sangyōshōhen, *Shōkō seisakushi, gyōsei kikō*, vol. 17, Tokyo: Shōkō Seisakushi Kankōkai, 1962, p. 19.

56 Tsuchiya Takeo, *Zoku Nihon keizaishi gaiyō*, Tokyo: Iwanami, 1941, p. 122ff. The Sapporo factory would become a leading center for machine manufacture, Akabane would be appropriated by the navy in 1883.

57 The government's other ironworks and mine, Nakakosaka, was similarly sold to a private concern, the Sakamoto Company, in 1884.

58 The word 'modern' with its implications of not only 'new' but 'superior' was not part of the Japanese lexicon in 1873. Words that I have loosely translated as 'modern,' that were most frequently used to describe Western machinery and the like, are *shinpo*, which meant 'progressing or advancing in civilization,' or prefaced by the characters meaning 'new' and 'good.'

59 These pronunciations/readings are based on the character's Japanese pronunciation or *kunyomi*. The *onyomi* or Chinese reading for both characters is *ki*. The Japanese word *kikai*, machine, uses the Chinese pronunciation. See *OHHS*, vol. 1, section 1, p. 7.

60 See for example Hilary Conroy, *The Japanese Seizure of Korea: 1868–1910, A Study of Realism and Idealism in International Relations*, Philadelphia, PA: University of Pennsylvania Press, 1960.

61 For a discussion of the Korean issue as a matter of national security see, Peter Duus, *The Abacus and the Sword: The Japanese Penetration of Korea, 1895–1901*, Berkeley, CA: University of California Press, 1995, pp. 48–60.

62 As demonstrated in Kenneth Pyle's study of mid-Meiji identity formation (*The New Generation in Meiji Japan: Problems of Cultural Identity, 1885–1895*, Stanford, CA: Stanford University Press, 1969), there was no consensus on how Japan defined its cultural identity during this period. While some, such as Tokutomi Sohō, sought to abandon everything Japanese in favor of total Westernization with the belief that unilinear historical progress would eventually merge all cultures into a single entity, others like Shiga Shigetaka urged a more moderate course (pp. 42–55). Although favoring the importation of Western artifacts and ideas, Shiga argued that there was beauty and value in Japanese culture and tradition that needed to be preserved. Shiga, like many in his day, believed that Japan should, in an effort to maintain its independence, become an industrial and commercial power through the adoption of the technological artifacts of the West – while upholding the essence of Japanese culture (pp. 57–8). Opposing any attempt to Westernize Japan were men like Motoda Eifu, who led the conservative vanguard which rejected the West and all it stood for. See Donald Shively, 'Motooda Eifu: Confucian Lecturer to the Meiji Emperor,' in Nivison, David S. and Arthur F. Wright, eds, *Confucianism in Action*, Stanford, CA: Stanford University Press, 1959, pp. 302–33. For other Confucian-based conservative reactions, see Donald Shively, 'Nishimura Shigeki: a Confucian View of Modernization,' in

Marius B. Jansen, ed., *Changing Japanese Attitudes Toward Modernization*, Rutland, VT: Charles E. Tuttle, 1982, pp. 193–241.

63 Ōkuma Shigenobu, '*Kan'yū no tame setchi shitaru kōgyō haraisage no gi*,' in Nihon Shiseki Kyōkai, eds., *Ōkuma Shigenobu kankei monjo*, vol. 4, Tokyo: (1935) 1985, p. 115.

64 Tokutomi Sohō, *Kōshaku Matsukata Masayoshi den*, vol. 1, Tokyo: Kōshaku Matsukata Masayoshi Denki Hakkōjo, 1935, p. 810.

65 Throughout the last years of the 1870s, Japan's economic situation steadily worsened to the point of major financial crisis. The government had followed a policy of deficit spending, in part to fund its efforts at 'civilization building.' To compensate, they issued inconvertible paper currency against their specie reserves, which lost both popular confidence and market value in a very short period of time. By 1878, people began hoarding silver, which only added to the problem, forcing interest rates higher and the value of the paper currency lower. Moreover, foreign trade took its toll on the government's dwindling specie reserves.

66 For the text of Ōkuma's loan proposal see, Inoue Kaoru kō Denki Hensankai, *Seigai Inoueko den*, vol. 3, Tokyo: Naigai Shoseki, 1933–34, pp. 144–6. Also Thomas C. Smith, *Political Change and Industrial Development in Japan: Government Enterprise, 1868–1880*, Stanford, CA: Stanford University Press, 1955, pp. 97–8.

67 For the text of Ōkuma's proposal see, Nihon Shiseki Kyōkai, eds., *Ōkuma Shigenobu kankei monjo*, vol. 4, Tokyo: Tokyo Daigaku Shuppankai (1935) 1985, pp. 113–24 (hereafter) *OSKM*.

68 *OSKM*, p. 114. It is debatable whether or not these facilities were profitable and in good order as Ōkuma claimed.

69 *OSKM*, p. 115.

70 *OSKM*, p. 115.

71 *OSKM*, pp. 113–14.

72 *OSKM*, p. 114. Ōkuma also identified the post office and telegraph, which he stated were 'not purely factories' as 'enterprises whose responsibility does not lay with the people.'

73 *Seigai Inouekō den*, 3: 149.

74 *SID*, 3: 149.

75 For contents of 'Regulations' see, Ōkurashō, *Meiji zaiseishi*, vol. 12, Tokyo: Ōkurashōnai Meiji Zaiseishi Hensankai Hensan, 1946.

76 Thomas C. Smith argues that the disposal of government enterprises was done for reasons of economic necessity. While this largely seems the case, I believe it is more complicated than what the data available to Smith at the time indicated. He and others all point out that the government did not intend to sell 'strategic' industries, as Ōkuma's and Inoue's proposals would indicate. However, no one has ever questioned why the government's iron industries, considered 'strategic' at the time, were sold. See Smith, *Political Change*, pp. 92–100.

77 'Yamao kōbu shōyū kōzan no gi ni tsuku kengen,' *Kōbun roku*, 2A–25–678 (4 January 1872) n.p.

78 'Yōkō seitetsu kikai setchu no gi ni tsuku,' *Kōbun roku: Kōbushō no bu*, 2A–25–1214–1 (March, 1874) n.p.

79 *KEH*, p. 25.

80 *KEH*, (emphasis mine).

81 Shively, 'Nishimura Shigeki,' p. 199.

82 Kenneth Pyle, 'Meiji Conservatism,' in Marius B. Jansen, ed., *The Cambridge History of Japan*. Vol. 5, Cambridge: Cambridge University Press, 1989, p. 695.

83 It is difficult to identify an exact date, or even year, for the 'end' of the so-called movement to attain 'civilization and enlightenment.' Part of this has to do with the government's contradictory policies regarding the adoption of Western icons and systems. At the same time that Japan was, for example, beginning to assert itself as an

imperialist power in Korea, the government was holding gaudy costume balls at the Rokumeikan, a Western-style gentleman's club that many considered the ultimate image of slavery to Westernization. See for example, Shively, 'Nishimura Shigeki,' p. 213. There were, however, clear trends against Westernization beginning in the mid- to late-1870s, the period traditionally identified as the early heyday of *bunmei kaika*. I would argue that this issue is confounded by the many ways in which intellectuals and government officials sought to redefine Japan's identity within the nexus of a rapidly changing political sphere.

84 From the day officials announced their intentions to build an ironworks at Kamaishi until late 1880, there was almost no coverage of the events in Iwate prefecture. In a survey of the *Tokyo nichi nichi shinbun*, *Yūbin hōchi shinbun*, and the *Chōya shinbun* from 1874 through 1880, there appear to be only four or five articles that mention the government's iron mines or ironworks. Those that do mention Kamaishi, typically do so indirectly, e.g., 'Mr. Under Secretary Hasegawa of the Public Works Department, left the capital on the 7th instant for a tour of inspection of the Anin, Innai, and Kamaishi mines.' *Chōya shinbun*, (7 September 1880), n.p. (Translation based on 'Japanese News–Court, Political, Official,' *Japan Weekly Mail*, (11 September 1880): 1180.) There is also an article on Kamaishi village which appeared in the *Japan Weekly Mail* in 1875; 'Kamaishi,' *Japan Weekly Mail*, (28 August 1875): 740–2.

85 'Hakuraihin ni otoranu shihin,' *Yūbin hōchi shinbun*, (15 December 1880) n.p. Ironically, on the day that this article appeared, there was a fire at Kamaishi's Kogawa charcoal-producing facility which destroyed five buildings and consumed its fuel supply. The ironworks was forced to shut down until March 1882.

86 The Western Enlightenment theories on which *bunmei kaika* ideology was based posited that national progress was based on a universal pattern of historical development. As nations moved further along the path toward 'civilization' they would become more similar, eventually fusing into a single cultural entity. See Pyle, 'Meiji Conservatism,' pp. 674–6.

87 Obana served in the Ministry of Public Works Mining Bureau (*Kōbushō kōzankyoku*) from 1883 until the Ministry of Public Works was disbanded in 1885. He studied metallurgy in Europe from 1879 to 1883 while on assignment from the ministry's engineering department. After the Ministry of Public Works was abolished, he took a position as an engineer at the Hiroshima Prefectural Office. It was in that capacity that he traveled to France. He subsequently served in the Ministry of Agriculture and Commerce, the Akita Mines, Sapporo Mines, and helped in constructing the government's Yawata Iron and Steel Works in 1896. He later served at Yawata as Director of the Pig Iron Section. He finished his career in academia, first as Professor of Metallurgy at Tokyo Imperial University's College of Engineering and then as Director of the Akita Mining College. Iseki Kōrō, *Dai nihon hakushiroku*, vol. 5, kōgaku hakushi, Tokyo: Hattensha, 1930, pp. 40–1 and p. 54.

88 Obana does not describe manufacturing in terms of production of consumer goods and his discussion of manufactured iron is only for Japan. 'Manufactured' refers to things such as iron plate, rod, rail, cable, and nails. Obana Fuyukichi, 'Seitetsu kensetsuron' in Ōkurashō, *Tekkō*, Tokyo: Ōkurashō, 1892, p. 26. Also available in Meiji Bunkō Shiryō Iinkai, *Meiji zenki sangyō hattatsushi shiryō*, vol. 70, no. 4, Tokyo: Meiji Bunkō Shiryō Iinkai, 1971; and Itō Hirobumi, *Hisho ruisan: jitsugyō, kōgyō shiryō*, Hiratsuka Atsushi, ed. Tokyo: Hisho Ruisan Kankōkai, 1936, pp. 75–88.

89 Obana, 'Seitetsu kensetsuron,' p. 29.

90 Obana, 'Seitetsu kensetsuron,' pp. 26–32.

91 Japan ranked above Italy in naval tonnage, but was well behind in miles of railroad. Unlike Italy, Japan did not even make the list for pig iron and steel production. 'Seitetsu kensetsuron,' pp. 40, 27–8 respectively.

92 Soeda Juichi, 'Seitetsusho no setsuritsu wa kyūmu nari,' in Ōkurashō, *Tekkō*, Tokyo: Ōkurashō, 1892, pp. 18–19.

93 Ōkurashō, *Tekkō*, pp. 19–23.

94 For Li Honzhang's political activities see for example, Jonathan Spence, *The Search for Modern China*, New York: Norton, 1990, pp. 218–21; Yen-p'ing Hao and Erh-min Wang, 'Changing Chinese Views of Western Relations, 1840–1895,' *Cambridge History of China*, vol. 10, part. 2, Cambridge: Cambridge University Press, 1980, pp. 197–9.

95 Personal correspondence, Commodore Robert Shufeldt to Senator Aaron A. Sargent, 1 January, 1882, cited in Paul H. Clyde, compiler, *United States Policy Toward China: Diplomatic and Public Documents, 1839–1939*, Durham, NC: Duke University Press, 1940, p. 163. Shufeldt characterized the Qing navy as thoroughly modern yet lacking adequately trained personnel. See also John L. Rawlinson, *China's Struggle for Naval Development, 1839–1895*, Cambridge, MA: Harvard University Press, 1967, pp. 68–81 for more on Chinese efforts to build a modern navy.

96 Duus, *Abacus and Sword*, p. 51. Until 1875 when the issue over the Sakhalin Islands was 'settled,' Japan viewed Russia as its 'chief hypothetical enemy' (p. 60, n. 73).

97 Duus, *Abacus and Sword*, p. 51.

98 Duus, *Abacus and Sword*, pp. 61–65.

99 Following Kenneth Pyle's definition of the term 'conservative,' I use it to describe those who were against the rapid reconfiguration of societies by individuals. Conservatives did not simply favor maintaining the status quo, they believed in the evolution of a society in which its particular 'customs and traditions' were embedded. Conservatives stood in opposition to liberals, the people who supported the ideals of the Enlightenment with beliefs in universality and scientific rationalism. See Pyle, 'Meiji Conservatism,' pp. 674–5.

100 The most glaring example of mid-Meiji cultural materialism was the opening of the Rokumeikan in 1883.

101 Carlton J. H. Hayes, *A Generation of Materialism, 1871–1900*, New York: Harper & Row, 1941, p. 202. Hayes uses the phrase 'era of Conservative supremacy' while discussing British domestic political trends, although it applies to Germany, Austria, Belgium, and Russia as well (pp. 202–4). He argues that there was a world-wide 'resurgence of national imperialism' from approximately the late 1870s through the turn of the century that was the product of tariff protectionism, socializing legislation, and a 'new national imperialism.' In the last 'three decades [of the nineteenth century] greater progress was made toward subjecting the world to European domination than had been made during the three previous centuries (p. 216).'

102 Hayes, *A Generation of Materialism*, p. 216.

103 The quoted phrase is from Kuga Katsunan whose editorial was describing the reasons why he named his newspaper *Nihon*. Kuga articulated the conservative position well. Referring to Japan's adoption of Western liberal theories, philosophy, science, technology, and industrial systems – in short, Western civilization – he argued that these things 'ought not to be adopted simply because they are Western; they ought to be adopted only if they can contribute to Japan's welfare' (quoted in Pyle, *New Generation*, pp. 94–5). The spirit of selective adoption which Kuga expressed in 1889 is also particularly appropriate to this earlier period.

104 Jiro Kumagai notes a turning point in Fukuzawa Yukichi's writings at this time. Fukuzawa quickly came to the realization that, following the example of Western imperialists, military might was more important than national wealth. Kumagai, 'Enlightenment and Economic Thought,' pp. 26–7.

105 If one of the reasons for developing the iron industry was self-sufficiency and the ability to produce the weapons of war, it stands to reason that having an iron industry and thus a powerful military were a means of national independence.

106 For government expenditure on Kamaishi see *KEH*, pp. 431–54. Figures for 1870–77 were not kept for individual enterprises and are only given in lump sums. Total expenditure on Kamaishi is stated to have been ¥2,376,625; *Shōkō seisakushi: gyōsei kikō*, vol. 17, p. 19.

107 Matsukata Masayoshi, 'Kokka fukyō no konpon o shoreishi, fukyū no hi o habuju beki no ikensho,' (1873) in Waseda Daigaku Shakai Kagaku Kenkyūjo, eds., *Ōkuma bunsho*, vol. 2, Tokyo: Waseda Daigaku Shakai Kagaku Kenkyūjo, 1959, pp. 1–3.

108 Matsukata Masayoshi, *Nōsho hensan no gi*, in Meiji Bunken Shiryō Kankōkai, *Meiji zenki zaisei keizai shiryō shūsei*, vol. 1, Tokyo: Meiji Bunken Shiryō Kankōkai, 1962, pp. 520–2 (hereafter *MZZKSS*); Matsukata Masayoshi, *Kannō yōshi*, *MZZKSS*, 1: 522–30. Also see Matsukata Masayoshi, 'Kiito seizō no gi ni tsuki ukagai,' in *Ōkuma bunsho*, vol. 2, pp. 239–46.

109 Matsukata, *Kannō yōshi*, p. 523.

110 See Matukata Masayoshi, *Zaisei kanki gairyaku*, (1880) in *MZZKSS*, 1: 532–5.

111 Matsukata Masayoshi, 'Seihei seiri shimatsu gi' quoted in Inukai Ichirō, 'The Kōgyō Iken: Japan's Ten Year Plan, 1884,' *KSU Economics and Business Review*, 6 (May, 1979), p. 42.

112 Inukai, 'The Kōgyō Iken,' p. 71. The largest government subsidy went to improving transportation and communications, including marine transportation. These areas were considered crucial for expanding Japan's foreign trade, not to mention military capabilities, although the latter was never explicitly mentioned. The bulk of industrial subsidy was directed at what Inukai calls 'productive enterprises, including small-scale farming' (p. 70). Matsukata's idea with *Kōgyō iken* was not to provide financial assistance to private enterprise. Rather, he sought to provide a more favorable environment for Japan's agricultural industries because he believed they would benefit the nation's economy first and foremost.

113 Inukai, 'The Kōgyō Iken,' p. 69.

114 Ishizuka Hiromichi, 'Shokusan kōgyō seisaku no tenkai,' in Kajinishi Mitsuhaya, ed., *Nihon keizaishi taikei*, vol. 5, Tokyo: Tokyo Daigaku Shuppankai, 1967, pp. 60–1. Between 1874 and 1887 the Ministry of Home Affairs and the Ministry of Agriculture and Commerce spent nearly one million yen on livestock development (pp. 52–3, 60–1).

115 Matsukata identified 21 strategic products for support. These included: raw silk, tea, tobacco, rice, rapeseed oil, camphor, vegetable wax, Japanese paper, laquer ware, marine products, metal products, cotton and flax spinning, weaving, sake brewing, leather and ceramic production, glass, sulfuric acid, and soda manufacture. All, except for the last three items, were considered appropriate for small-scale production as agricultural by-employments. Maeda Masana, *Kōgyō iken*, in *MZZKSS*, 18: 436 *ff*.

116 See Ericson, *Sound of the Whistle*, for a discussion of Matsukata's proposals as Minister of Home Affairs for the privatization of Japan's railroads, especially pp. 97–122.

117 James R. Bartholomew, *The Formation of Science in Japan*, New Haven, CT: Yale University Press, 1989, pp. 131–2.

118 Inukai Ichirō notes that *Kōgyō iken* has been alternative translated as 'Advice for the Encouragement of Industry,' 'Economic White Book,' 'My View on Industry,' and 'Views on Industry.' Inukai prefers to call the 30-volume study 'A Proposal for Economic Development' or 'The Ten-Year Development Plan of 1884.' Inukai, 'Kōgyō Iken,' p. 4.

119 For most scholars of *Kōgyō iken*, the most significant issue is the difference between the Maeda's original version and the official version that was approved by Matsukata and the *Dajōkan*. The official text is missing Maeda's proposal for an industrial bank, at item which Matsukata strongly opposed.

120 *Kōgyō iken*, *MZZKSS*, 18: 45.

121 *Kōgyō iken*, *MZZKSS*, 18: 45.

122 Inukai states that 'there is ample evidence in the *Kōgyō Iken* that their hearts were in battleships and big factories.' Inukai, 'Kōgyō Iken,' p. 59.

123 Based on Bureau of Mines, *Mining in Japan, Past and Present*, Tokyo: The Bureau of Mines, The Department of Agriculture and Commerce of Japan, 1909, p. 57.

124 Bureau of Mines, *Mining in Japan*, p. 57.

125 'Ryūshitsu kōmyaku ni atarazu,' *Jiji shinbun*, (10 April 1882) n.p. The report issued to the Ministry of Public Works which condemned Kamaishi for other reasons clearly stated that the iron ore at Kamaishi was of very high quality. See Saigusa Hiroto and Iida Ken'ichi, eds. *Nihon kindai seitetsu gijutsu hattatsushi: Yawata seitetsujo no kakuritsu katei*, Tokyo: Tōyō Keizai Shinbunsha, 1957, p. 65–6.

126 As discussed in Chapter 4, Kamaishi was temporarily shutdown because of human error (an accidental fire which destroyed the fuel supply).

127 Nihon Tekkōshi Hensankai, eds., *Nihon tekkōshi, Meiji-hen*, Tokyo: Gogetsu Shobō, (1945) 1982, p. 111.

128 The full text of Itō's report is in Saigusa and Iida, *Nihon kindai seitetsu*, p. 65–6.

129 Asano Sōichirō purchased his factory with Shibusawa Eiichi's help and subsequently changed its name to the Asano Cement Company. Later it became Nihon Cement Company, Ltd. He would go on to make a fortune in coal and coke, forming a *zaibatsu* (conglomerate) of the same name.

130 Nihon Tekkōshi Hensankai, *Nihon tekkōshi*, p. 111. The details of Asano's attempts to re-spark interest in Kamaishi are based on excerpts from his diary.

131 Kobayashi Masaaki, *Nihon no kōgyōka to kangyō haraisage: seifu to kigyō*, Tokyo: Tōyō Keizai Shinpōsha, 1977, p. 212.

132 Nihon Tekkōshi Hensankai, *Nihon tekkōshi*, p. 112.

133 Nihon Tekkōshi Hensankai, *Nihon tekkōshi*, p. 113; Kobayashi, *Nihon no kōgyōka*, p. 211.

134 Nihon Tekkōshi Hensankai, *Nihon tekkōshi*, p. 114.

135 Nihon Tekkōshi Hensankai, *Nihon tekkōshi*, p. 115.

136 Nihon Tekkōshi Hensankai, *Nihon tekkōshi*, p. 115.

137 Yonekura Seiichiro, *The Japanese Iron and Steel Industry, 1850-1990: Continuity and Discontinuity*, New York: St. Martin's Press, 1994, p. 31.

138 Four of the former domains, Satsuma, Chōshū, Hizen, and Tosa, had played a key role in the Meiji Restoration. Sat-Chō alliance, as well as *hanbatsu* (clan-faction) politics, refers to Meiji officials from Satsuma and Chōshū who dominated the government, especially during the 1880s. Their power lasted well into the 1920s.

139 Joyce Chapman Lebra, 'Ōkuma Shigenobu and the 1881 Political Crisis,' *Journal of Asian Studies*, 18: 4 (August 1959): pp. 482–3. Lebra notes that historians disagree on the price which Godai offered for the land. Figures range from Lebra's ¥300,000 to ¥387,082. The actual value of the property was estimated at ¥21,000,000, and the government had spent more than ¥14,000,000 on the development project. See also Borton, *Japan's Modern Century*, pp. 122–4.

140 According to Lebra, the major progressive newspapers, *Tokyo-Yokohama mainichi shinbun*, *Yūbin hōchi shinbun*, and *Chōya shinbun* carried severe criticism of the sale. The *Tokyo nichi nichi shinbun*, a conservative semi-official paper, was equally vocal in its disapproval of the proposed sale; Lebra, p. 483.

141 The political crisis of 1881 was over much more than land. In fact, the land deal can be seen as a mirror of the political struggles between Ōkuma, Itō, and the other members of the Sat-Chō alliance regarding the problems surrounding the establishment of a national Diet and constitution in the face of growing government conservatism. Although Ōkuma was forced out of government, public outcry forced the bureaucracy to announce their intentions to formulate a constitution and convene a national representative assembly at a significantly earlier date than was originally planned.

142 Nihon Tekkōshi Hensankai, *Nihon tekkōshi*, p. 111.

143 *KEH*, p. 136.

144 Hyakunenshi Hensan Iinkai, *Tetsu to tomo ni hyakunen*, vol. 1, Kamaishi City, Iwate: Shin Nihon Seitetsu Kabushikigaisha, 1986, p. 55 (hereafter *TTH*).

145 Fuji Seitetsu Kabushikigaisha Kamaishi Seitetsujo, *Kamaishi seitetsujo shichijū nenshi*, Kamaishi: Fujiseitetsu Kabushikigaisha Kamaishi Seitetsujo, 1955, p. 42 (hereafter *FSK*). Tanaka paid ¥12,600 for Kamaishi.

146 Nihon Tekkōshi Hensankai, *Nihon tekkōshi*, p. 115.
147 I say purportedly because the Ikuno and Sado mines, along with the Takashima and Miike coal mines, were all sold at substantial profit. See Kobayashi, *Nihon no kōgyōka*, pp. 138–9.
148 Kobayashi, *Nihon no kōgyōka*, p. 138. Based on an average of the percent of investment as reflected in sales price, not including the enterprises which the government sold at a profit.
149 *FSK*, p. 43.
150 Iida Ken'ichi, *Nihon tekkō gijutsushiron*, Tokyo: Mihari Shobō, 1973, pp. 417–39. The other men were Saigō Takamori and Ijūin Kanetsugu.
151 The text of Tanaka's letter is reprinted in *TTH*, 2: 369–70; and *FSK*, pp. 46–8.
152 Pyle observes that the early 1890s were a time when Meiji intellectuals were struggling with identity formation – or perhaps more appropriately re-formation. Men like Kuga and Tokutomi were wrestling with the contradictions of how to preserve Japan's history and the new national consciousness in the face of the previous three decades of Westernization: Pyle, *New Generation*, pp. 162–87.
153 See for example, 'The Study of English by Japanese,' *Japan Weekly Mail*, (25 November 1871): 656–8, which argues that the Japanese need to study English if they want to stop imitating and start understanding what it means to be 'civilized'; 'Unsound Progress,' *Japan Weekly Mail*, (30 December 1871): 719–21, which sees Japanese ideas of progress as unflattering imitation; 'The Return of the Embassy,' *Japan Weekly Mail*, (23 September 1873): 594–6, which doubts that any policy changes will result from the Iwakura Mission because its members could only gain a superficial understanding of the West; 'The Return of Japanese Students From Abroad,' *Japan Weekly Mail*, (19 June 1874): 380–1, that praises the spirit behind the government's sending of students to Europe, but doubts that the effort will produce any positive results; a letter in response to 'The Return of Japanese Students From Abroad,' *Japan Weekly Mail*, (8 July 1874): 489, which, in a paternalistic tone, takes aim at the former students and hopes that the government will see the error of its ways. Also n. 11 above. For Japanese criticisms see 'Japanese Dress for the Japanese' *Yūbin hōchi shinbun*, April 1878 (reprinted in *Japan Times*, 6 April 1878, pp. 292–3); 'On Abandoning the Use of Leathern Boots for Japanese Soldiers,' *Yūbin hōchi shinbun*, March 1878 (reprinted in *Japan Weekly Mail*, 2 February 1878, pp. 95–6).
154 Ironically, Westerners who first criticized the Japanese for giving up traditional clothing, and then later rejoiced in seeing the return of the kimono to Tokyo's formal events, did not realize that the reappearance of the preferred 'traditional garb' was part of a growing anti-Westernism in Japan. See Sally Hastings, 'The Empress' New Clothes and Japanese Women, 1868–1912,' *The Historian*, 55:4 (Summer 1993): 677; and Shibusawa Keizo, ed. *Japanese Life and Culture in the Meiji Era*, Charles S. Terry, trans., Tokyo: Pan-Pacific Press, 1958, p. 27.
155 Kawase Hideharu, 'Kangyōron,' (December, 1878) in *Ōkuma bunsho*, 2: 278.
156 *Ōkuma bunsho*, 2: 278.
157 *Ōkuma bunsho*, 2: 280–1.
158 In *Jiji shōgen* (A Brief Discussion of Current Affairs, 1881) Fukuzawa abandoned his support of popular rights in favor of creating a unified government. Fukuzawa's new position was in reaction to the political crisis of 1881; see *FYZ*, 5: 98. This is often considered the turning point in Fukuzawa's philosophy. See, Kumagai, 'Enlightenment and Economic Thought,' p. 27. Fukuzawa continued to harden his stance throughout the 1880s, first in *Gaikō-ron* (On Diplomacy, 1883) and later in *Datsua-ron* (Going Out of Asia, 1885) where he advocated Japanese intervention in Chinese and Korean domestic affairs in the name of 'progress' and 'national security.' See *FYZ* 9: 195–6 and 10: 240, respectively. It should be pointed out that by and large, Fukuzawa's 'enlightenment thought' may best be described as scientific rationalism.

159 *FYZ*, 5: 118–20. I have placed 'anti-Western' in quotes because there are frequent contradictions is Fukuzawa's arguments. Fukuzawa would reject Christianity, for example, as Western meddling with Japanese minds, although he had previously supported the spread of Western religions (and philosophies) in Japan. He would change his position on Christianity again in 1884, stating that Japan could not go wrong in following Western religions (*FYZ*, 7: 212–13). He similarly spoke of 'civilizing' Asia to help fend off the Western challenge, only to further elaborate this position by arguing that Japan should take over Asia before the West did. Although not part of the 'new generation' and Pyle's study, Fukuzawa was clearly an intellectual in crisis.

160 Ōkuma complained bitterly about the West refusing to acknowledge Japan's 'progress toward civilization.' A decade later Tokutomi Sohō would similarly attack Western commentators on this account. See Pyle, *New Generation*, pp. 166–8.

161 Carl von Clausewitz, Michael Howard, and Peter Paret, eds. Trans., *On War*, Princeton, NJ: Princeton University Press, 1984, p. 87.

Conclusion

1 The adage 'know your enemy' can be traced to the Chinese *Sun Tzu* as well as Miyamoto Musashi's *Book of Five Rings*. In his *New Thesis of 1825* (*Shinron*) Aizawa Seishisai also talked of the importance of knowing one's enemy before being able to launch on an effective counter action; see Bob Tadashi Wakabayashi, *Anti-Foreignism and Western Learning in Early-Modern Japan: The New Thesis of 1825*, Cambridge, MA: Council on East Asian Studies, Harvard University, 1991, pp. 248–50.

2 Ōkubo, for example, believed that the reasons for British wealth and power were the development of the iron industry, export markets, and moreover, the attention payed to the transportation network. See Nakamura Masanori, Ishii Kanji, and Kasuga Yutaka, eds, *Nihon kindai shisō taikei*, vol. 8, *keizai kōsō*, Tokyo: Iwanami Shoten, 1988, p. 244.

3 This well-known phrase is from the Charter Oath of 1868; Tsunoda, Ryusaku, Wm. Theodore De Bary, and Donald Keene, *Sources of Japanese Tradition*, vol. 2, New York: Columbia University Press, 1964, p. 137.

4 If one defines a technological artifact as a tool with which man shapes, or has shaped, his environment, it is also possible to include things such as the Meiji absorption of French legal codes within a general discussion of technological change.

5 See Fukuzawa Yukichi's *Bunmeiron no gairyaku*, in Fukuzawa Yukichi, *An Outline of a Theory of Civilization*, David A. Dilworth and G. Cameron Hurst, trans., Tokyo: Sophia University Press, 1973, p. 37.

6 Even those internalist commentators who called for the Westernization of Japanese education or the elimination of certain 'evil customs' were not looking to erase Japan's history. By urging that the Japanese adopt the enlightenment spirit they hoped to catapult the country back onto the proper path. Fukuzawa, for example, argued that centuries of bowing to authority had stifled the Japanese spirit. Once politically and spiritually liberated, however, even commoners could contribute to the national cause. Fukuzawa, *An Outline of a Theory*, pp. 16–17.

7 I am of course referring to *kokutai*, Japan's national polity which embraced the emperor as the center of a political system; although based on Western models, it emphasized a distinct Japanese culture and morality.

8 Kenneth B. Pyle, *The New Generation in Meiji Japan: Problems of Cultural Identity, 1885–1895*, Stanford, CA: Stanford University Press, 1969, p. 118.

9 Throughout his career, Ōkuma Shigenobu complained that British and French officials refused to recognize Japan's 'progress toward civilization.' These foreign officials only complained about Japan's backwardness. As official efforts to renegotiate the unequal treaties intensified during the early- to mid-1890s, many

Japanese intellectuals like Tokutomi Sohō were faced with the dilemma of reconciling their past calls for the Westernization of Japan with the West's refusal to acknowledge Japan's true progress in achieving a higher 'level of civilization.' See Pyle, *New Generation*, p. 161–7.

10   I define *techno-imperialism* as the use of modern military technologies to facilitate territorial acquisition in an effort to find both the raw materials which support, and markets for, the aggressor's industries.

11   I do not intend to imply that Japanese imperialism was based solely on the need to acquire iron ore (or coal) to support its navy. Japan's moves into Manchuria and Korea were, however, fueled by the navy's demand for large amounts of iron and steel. I would argue that this is an issue which needs to be more closely examined when looking at the factors which contributed to Japanese imperialism. The 'host of other factors' include: protecting previous economic investments in Korea and Manchuria, attempting to limit Russia's southward expansion, acquisition of raw materials, and securing a sphere of political influence in East Asia in the name of national security.

12   Most of the literature on the subject of Meiji industrialization discusses the process in this light. See for example, Smith (1948), who argues that government initiative was instrumental in Japan's economic development; Eleanor D. Westney, *Imitation and Innovation: The Transfer of Western Organizational Patterns to Meiji Japan*, Cambridge, MA: Harvard University Press, 1987, who argues that political needs were the basis of an ad hoc industrialization 'policy'; or Yasuzō Horie, 'Modern Entrepreneurship in Meiji Japan' in William W. Lockwood ed., *State and Economic Enterprise in Japan*, Princeton, NJ: Princeton University Press, 1965, pp. 183–203, who states that military parity and treaty revision were the goals of Meiji industrial modernization. Tessa Morris-Suzuki, *The Technological Transformation of Japan: from the Seventeenth to the Twenty-first Century*, Cambridge: Cambridge University Press, 1994, argues that while state and local initiatives were important, the overarching factor was a social network that facilitated the absorption and expansion of technology. Masayuki Tanimoto, *The Role of Tradition in Japan's Industrialization*, Oxford: Oxford University Press, 2006, demonstrates that traditional industries working in small, Tokugawa-era workshops contributed to the Meiji industrialization.

13   For example see, Nakamura, *Nihon kindai shisō taikei*, pp. 241–75. Nakamura argues that Japanese industrialization was not simply a government or state-led initiative. It involved significant participation on the local level and that there is too much importance placed on *fukoku* theory and Japan's so-called catch-up mentality.

14   For criticism of Japan's new eating habits, Western dress, and other 'frivolous' artifacts, see Kawamura Yoshia, 'Bunmei kaika ron,' *Tokyo Nichinichi Shinbun*, no. 1756, (8 October 1877): 931–2. For later examples of rejected artifacts or ones that were losing popularity, e.g., Western-style clothes, especially for women, see Sally Hastings, 'The Empress' New Clothes and Japanese Women, 1868–1912, *The Historian*, 55: 4 (Summer 1993) 678.

15   Kawamura, 'Bunmei kaika ron,' p. 931.

16   For Tokutomi and changing attitudes toward the West, see Pyle, *New Generation*, pp. 144–87. Fukuzawa's writings of the late 1870s and early to mid-1880s illustrate his dilemma. Within a few short years, he argued in favor of popular rights and then for restricting popular rights, for equality among nations and for Japan's need to become a military power, asserting its authority in Asia. See for example, Tomita Masafumi and Tsuchibayashi Shunichi, eds, *Fukuzawa Yukichi zenshū*, Tokyo: Iwanami Shoten, 1959, 5: 108, 7: 259, and 9: 195–6.

17   Shinmachi was sold to Mitsui in 1887, six years before Tomioka; Kobayashi Masaaki, *Nihon no kōgyōka to kangyō haraisage: seifu to kigyō*, Tokyo: Tōyō Keizai Shinpōsha, 1977, p. 138. Shinmachi's opening ceremony was attended by a number of Meiji officials including Ōkubo, Ōkuma, Itō, Matsukata, and Sano. Also in attendance were a German adviser named Bayer and a Swiss named Bell who had helped with

construction of the facility. In his report on the ceremony, Gunma prefectural official Sayatori Motohiko stated that Tomioka had brought civilization to Gunma prefecture and that the new waste thread facility would do the same. He continued that technical progress in silk reeling is based on Japanese intelligence and that it would demonstrate Japan's progress toward civilization; Sayatori Motohiko, 'Kōko sōdan, *Tokyo nichinichi shinbun*, no. 1769 (23 October 1877): 982

18 Yonekura calls Kamaishi 'the most eligible candidate among them,' i.e., the government's failed industries which Matsukata targeted for disposal; Yonekura Seiichiro, *The Japanese Iron and Steel Industry, 1850-1990: Continuity and Discontinuity*. New York: St. Martin's Press, 1994, p. 27.

19 Even reports to the international exhibitions were very sketchy on the details of the government's ironworks; *International Exhibition of 1876: Official Catalogue of the Japanese Section, and Descriptive notes on the Industry and Agriculture of Japan*, Philadelphia, PA: Japanese Commission, 1876, p. 48. Contemporary Western engineering reports discuss Kamaishi in limited terms: equipment, equipment manufacturers, foreign employees, resources, and market prices. The tone of some of the earlier articles, from 1875 and 1876, could be seen as approving of Japanese 'progress.' Later reports are largely technical in nature. This could be because the earlier articles were written by Kamaishi's designer, David Forbes. See David Forbes, 'Report on the Progress of the Iron and Steel Industries in Foreign Countries,' *Journal of the Iron and Steel Institute*, (1875, part 1); David Forbes, 'Report on the Progress of the Iron and Steel Industries in Foreign Countries,' *Journal of the Iron and Steel Institute*, (1875, part 2); and 'Report on the Iron and Steel Industries of Foreign Countries: Japan,' *Journal of the Iron and Steel Institute*, 1880: 1, pp. 381–3.

Appendix I

1 This translation is based on the proposals as they appear in Ōshima Shinzō, *Ōshima Takatō kōjitsu* (Hyōgō: Ōshima Shinzō, 1937), 781–4; Fuji Seitetsu Kabushikigaisha Kamaishi Seitetsujo, *Kamaishi seitetsujo shichijū nenshi* (Kamaishi: Fujiseitetsu Kabushikigaisha Kamaishi Seitetsujo, 1955), 20–1; Iida Kenichi, *Gendai nihon no gijutsu to shisō* (Tokyo: Toyo Keizai Shinpōsha, 1974), 208–11; Kobayashi Masaaki, *Yawata seitetsujo* (Tokyo: Hanbai Kyoikusha Shuppan Sabisu, 1972), 101-2; Saigusa Hiroto and Iida Ken'ichi, eds, *Nihon kindai seitetsu gijutsu hattatsushi: Yawata seitetsujo no kakuritsu katei* (Tokyo: Tōyō Keisai Shinpōsha, 1957), 31–4; and on the translation in David G. Wittner, *Technology Transfer in the Meiji era: The Case of Kamaishi Iron Works, 1874–1890*, Master's Thesis (Columbus, OH: The Ohio State University, 1995), 28–9. I have synthesized the extant versions to give the most comprehensive view of Ōshima's and Bianchi's proposals, i.e., include data present in one proposal not present in another, but have maintained the traditional structure and order of the original.

2 There were 5 families on farms at Suzuko and 20 families and well-developed farmlands at Otadagoe. The expense for relocation of the Suzuko families was ¥248. Figures for Otadagoe are unavailable, but it is reasonable to assume that expenses would have been greater given the increased numbers and acreage (see Kobayashi, *Yawata seitetsujo*, p. 105, for relocation costs).

Appendix II

1 Nōshōmushō Nōmukyoku, *Daiichiji zenkoku seishikōjo chōsahyō* (Tokyo: Nōshōmushō Nōmukyoku, 1895).

2 One *kin* is approximately 1.33 pounds.

Bibliography

Archive and manuscript collections

Akte: OBA 99, Technische Universität Bergakademie Freiberg, Freiberg, Germany.
Dajō ruiten, Kokuritsu kōbunshokan, Tokyo.
Foreign Office. General Correspondence, Japan. (F.O. 46).
Kōbun roku, Kokuritsu kōbunshokan, Tokyo, and Shiryōhensanjo, University of Tokyo.
Ōkuma Monjo, Waseda University, Tokyo.

Published manuscript and collections

Asakura Harukiko, ed. *Meiji shoki kan'in mokuroku: shokuinroku shūsei*. 4 vols. Tokyo: Kashiwa Shobō, 1981.
Commercial Reports from Her Majesty's Consuls in Japan, London: Harrison and Sons, 1869–1880.
Gunma Kenritsu Bunshokan. *Gunma kenshi shūshū fukusei shiryō mokuroku*. 4 vols. Maebashi: Gunma Kenritsu Monjokan, 1994.
Inoue Kaoru. 'Itō-kō Inoue-kō shokan' *Inoue Kaoru kankei monjo* 112, n.d., n.p. (Letters to Shibusawa and others from Inoue and Itō, ca.1870–1872) University of Tokyo, Faculty of Economics.
Inoue Kaoru kō Denki Hensankai. *Ōkurashō jidai no Inoue kō*. 10 vols. Tokyo: Inoue Kaoru kō Denki Hensankai, 1928–1933 (University of Tokyo, Faculty of Economics).
Nihon Kōgyōshi Shiryōshū Hensan Iinkai, ed. 'Watanabe Wataru, Firudo no-to (1877),' *Nihon kōgyō shiryōshū*, Meiji-hen, 16, mae, no. 1, Tokyo: Nihon Kōgyōshi Shiryō Kankō Iinkai, 1992.
Nihon Kōgyōshi Shiryōshū Hensan Iinkai, ed. 'Yamao Yōzō: shokanshū' *Nihon kōgyō shiryōshū*. Meiji-hen, 15, ato, no. 1, Tokyo: Nihon Kōgyōshi Shiryō Kankō Iinkai, 1991.

Newspapers and serials

Chōya shinbun, Tokyo, 1874–82.
Hakumon shinshi, Tokyo, 1872–73.
Jiji shinpo, Tokyo, 1880–82.
Minkan zasshi, vols 1–12, Tokyo, 1874–75.
Ryūmon zasshi, Tokyo: Ryūmonsha. 1890–95.
Shinbun zasshi, Tokyo: 1872–73.
Tokyo nichi nichi shinbun, Tokyo, 1874–95.

Yokohama shinbun shōsho, Yokohama, 1871–72.
Yūbin hōchi shinbun, Tokyo, 1874–82.
China Directory, Hong Kong: China Mail Office, 1865–73.
Chronicle and Directory for China, Japan, and the Philippines, Hong Kong: Daily Press Office, 1867–69.
Far East: an illustrated fortnightly newspaper, Yokohama, 1870–75.
Japan Gazette, Hong Lists and Directory, Yokohama: Office of the Japan Gazette, 1875–81.
Japan Punch, (reprint edition, 1870–73). Tokyo: Yūshōdō, 1975.
Japan Weekly Mail, Yokohama, 1871–96.
L'Echo du Japon, Yokohama, 1875–80.
Moniteur des Soies, Lyons, France, 1869–75.
Naniwa shinbun, 1875–77.

Japanese-language materials

Andō Yasuo. 'Tomioka seishijō.' Chihōshi Kenkyū Kyōgikai, ed. *Nihon sangyōshi taikei*. Vol. 4. Tokyo: Tokyo Daigaku Shuppankai, 1959.
Anzai Toshimitsu. *Fukuzawa Yukichi to seiō shisō*. Nagoya: Nagoya Daigaku Shuppankai, 1995.
Asahi Shinbunsha, ed. *Nihon rekishi jinbutsu jiten*. Tokyo: Asahi Shinbunsha, 1994.
Banno Junji. '"Fukoku' ron no Seijishiteki kōsatsu.' In Umemura Mataji and Nakamura Takafusa, eds, *Matsukata zaisei to shokusan kōgyō seisaku*. Tokyo: United Nations University Press, 1983.
Chigusa, Y. *Fukuzawa Yukichi no keizai shisō: sono gendaiteki igi*. Tokyo: Dōbunkan, 1994.
Chōya shinbun. (7 September 1880) n.p.
Edamatsu Shigeyuki, Sugiura Tadashi and Yagi Kosuke. *Meiji nyūsu jiten*. Tokyo: Mainichi Komyunikēnshonzu Shuppanbu, 1983.
Fuji Seitetsu Kabushikigaisha Kamaishi Seitetsujo. *Kamaishi seitetsujo shichijū nenshi*. Kamaishi: Fujiseitetsu Kabushikigaisha Kamaishi Seitetsujo, 1955.
Fukui Jun. *Nihon kōhō zenshū*. Osaka: Hōgyokudō, 1887.
Fukuzawa Yukichi. 'Gaikokujin no naichi zakkyo yurusu bekarazaru no ron.' *Minkan zasshi*, 6 (January, 1875): 6–7.
—— *Minkan keizairoku*. Vol. 2. Tokyo (December 1880).
Gaimushō Hensan. *Nihon gaikō bunsho*. Vol. 2. Tokyo: Gaimushō Shuppan, 1964.
Gunma Kenritsu Rekishi Hakubutsukan, eds. *Futatsu no seishi kōjo: Tomioka seishijo to Usuisha*. Takasaki: Gunma Kenritsu Rekishi Hakubutsukan, 1987.
Gunma Kenshi Hensan Iinkai, eds. *Gunma-kenshi*. 27 vols. Maebashi: Gunma-ken, 1877–82.
Gunma-ken Naimubu, *Gunmaken sanshigyō genkyō chōsasho*. 2 vols. Maebashi: Gunmaken Naimubu, 1904. Compiled by Meiji Bunken Shiryō Kankōkai. *Gunmaken sanshigyō genkyō chōsasho*. Tokyo: Meiji Bunken Shiryō Kankōkai, 1969.
'Hakuraihin ni otoranu shihin.' *Yūbin hōchi shinbun*. (15 December 1880) n.p.
Hanzawa Shūzō, *Nihon seitetsu kotohajime: Ōshima Takato no shōgai*. Tokyo: Shinjinbutsu Oraisha, 1974.
Harada Katsumasa. *Testsudō to gendaika*. Tokyo: Yoshikawa Kōbunkan, 1998.
Hayami Kenzō. 'Hayami Kenzō rireki bassui.' *Gunma-kenshi, shiryō-hen*, Gunma Kenshi Hensan Iinkai, eds. Vol. 23, Maebashi: Gunma-ken, 1985.
'Hōkoku: kōbushō go-kōnai seishijo.' *Tokyo nichi nichi shinbun*. No. 813 (2 October 1874).
'Hōkoku.' *Tokyo nichi nichi shinbun*. No. 771 (15 August 1874) n.p.

Hosokawa Junjirō. *Seishi hitsukei*. Nihonbashi: Yamashiroya, n.d..

Hyakunenshi Hensan Iinkai. *Tetsu to tomo ni hyakunen*. 2 vols. Kamaishi City, Iwate: Shin Nihon Seitetsu Kabushikigaisha, 1986.

Igarashi Akio. *Meiji isshin no shisō*. Kanagawa: Seori Shobō, 1996.

Iida Kenichi. *Nihon kagaku gijutsushi*. Tokyo: Asahi Shinbunsha, 1964.

——*Nihon tekkō gijutsushiron*. Tokyo: Mihari Shobō, 1973.

——*Gendai nihon no gijutsu to shisō*. Tokyo: Tōyō Keizai Shinpōsha, 1974.

——*Nihon tekkō gijutsu*. Tokyo: Tōyō Keizai Shinhō, 1979.

——*Nihon tekkō gijutsu no keisei to tenkai*. United Nations University Research Paper No. HSDRJE-8J/UNUP-28. Tokyo: United Nations University Press, 1979.

——*Tetsu no kataru nihon no rekishi*. Tokyo: Soshiete, 1979.

Iida Ken'ichi, ed. *Jukogyōkan no tenkai to mujun gijutsu no kaishashi*. Vol 4. Tokyo: Yuhikaku, 1982.

——*Tetsu no 100-nen Yawata Seitetsujo*. Tokyo: Daiichi Hoki, 1988.

——*Nihon kindai shisō taikei, kagaku to gijutsu*. Vol. 14. Tokyo: Iwanami Shoten, 1989.

Inoue Kaoru kō Denki Hensankai, eds. *Seigai Inoueko den*. 5 vols. Tokyo: Naigai Shoseki, 1933–34.

Iseki, Kōrō *Dai nihon hakushiroku*. Vol. 5, *kōgaku hakushi*. Tokyo: Hattensha, 1930.

Ishihara Kōichi. *Nihon gijutsu kyōiku shiron*. Tokyo: San'ichi Shobō, 1962.

Ishii Kendo. *Meiji jinbutsu kigen*. Vol. 6. Tokyo: Sakuma Gakugei Bunko (1908) 1997.

Ishizuka Hiromichi. *Nihon shihon shugi seiritsushi kenkyū*. Tokyo: Yoshikawa Bunken, 1973.

——'Shokusan kōgyō seisaku no tenkai.' *Nihon keizaishi taikei*. Vol. 5. Kajinishi Mitsuhaya, ed. Tokyo: Tokyo Daigaku Shuppankai, 1967.

Itō Hirobumi, *Itō-kō enzetsusho*. Tokyo: Tokyo Nichinichi Shinbun, 1899.

——*Itō-kō zenshū*. Edited by Komatsu Midori. Vol. 2. Tokyo: Itō-kō Zenshū Kankokai, 1927.

——*Hisho ruisan: jitsugyō, kōgyō shiryō*. Edited by Hiratsuka Atsushi. Tokyo: Hisho Ruisan Kankōkai, 1936.

Itō Hirobumi Kankei Monjo Kenkyūkai. *Itō Hirobumi kankei monjo*. Vol. 8. Tokyo: Hanawa Shobō, 1976–1980.

Itō Moemon. *Chiigai sanji yōroku*. Tokyo: Itō Moemon, 1886.

Itō Yajirō. 'Nihon seitetsuron.' *Kōdan zasshi*. 10 (26 February 1890): 27–30.

——'Nihon seitetsuron.' *Kōdan zasshi*. 11 (23 March 1890): 24–7.

——'Nihon seitetsuron.' *Kōdan zasshi*. 12 (23 April 1890): 25–7.

Iwakura Shōko, ed. *Iwakura shisetsudan to Itaria*. Kyoto: Kyoto Gakujutsu Shuppankai, 1997.

Iwanami Shoten Hensanbu. *Nihon kindai shisō taikei*. Vol. 1, *kindai shiryō kaisetsu, sōmokuroku, sakuin*. Tokyo: Iwanami Shoten, 1992.

Kada Naoji, *Isshin sangyō kensetsushi*. 1932.

'Kamaishi kōzan wa zenzan hotondo tetsu nari.' *Kōzan zasshi*. 3 (7 December 1893): 297–8.

'Kamaishi kōzan no saikō.' *Kōzan zasshi*. 3 (15 July 1893): 128–9.

Kamijō Hiroyuki. *Kinu hitosuji no seishun: 'Tomioka nikki' ni miru Nihon no kindai*. Tokyo: NHK Books, 1978.

——'Pooru Buryuna: kikai seishi gijutsu no dokusōteki ishokusha.' *Kōza Nihon no shakaishi*. Vol. 2. Tokyo: Nippon Heironsha, 1986.

Kaneko Kentarō, ed. *Itō Hirobumi den*. Tokyo: Shumpo Ko Tsuishōkai, 1940–44.

Kannōkyoku. *Kyū kangyōryō dai ikkai nenpō, Meiji 9-nen*. Tokyo: Kannōkyoku, 1877.

Kannōkyoku, Shomukyoku. *Kyōshinkai hōkoku, kenshi no bu*. Tokyo: Yūrindō, 1880.

Kawamura Yoshia. 'Bunmei kaika ron.' *Tokyo Nichinichi Shinbun*. no. 1756 (8 October 1877): 931–3.

Kawase Hideharu. 'Kangyōron.' (December, 1878). *Ōkuma bunsho*. Vol. 2. Waseda Daigaku Shakai Kagaku Kenkyūjo, eds. Tokyo: Waseda Daigaku Shakai Kagaku Kenkyūjo, 1959.

Kinugawa Taichi. *Honpō Menshi Bōsekishi*. Ōsaka: Nihon Mengyō Kurabu, 1937. Reprinted in *Meiji hyakunenshi sōsho*. Vol. 7. Tokyo: Hara Shobō, 1990.

Kitajima Masayoshi. *Seishigyō no tenkai to kōzō: Bakumatsu isshinka Suwa ni tsuite chōsa hōkoku*. Tokyo: Hanzawa Shobō, 1970.

Kobayashi Masaaki. *Yawata seitetsujo*. Tokyo: Hanbai Kyoikusha Shuppan Sabisu, 1972.

——*Nihon no kōgyōka to kangyō haraisage: seifu to kigyō*. Tokyo: Tōyō Keizai Shinpōsha, 1977.

Kojima Sei'ichi. *Nihon tekkōshi, Meiji-hen*. Tokyo: Sakura Shobō, 1945.

'Kōko sōdan.' *Tokyo nichinichi shinbun*. No. 603 (8 February 1874).

'Kōko sōdan.' *Tokyo nichinichi shinbun*. No. 836 (29 October 1874).

Kondō Kazuhiko. *Bunmei no hyōshō Eikoku*. Tokyo: Yamagawa Shuppansha, 1998.

Kōza Nihon gijustu no shakaishi. 10 vols. Tokyo: Nippon Heironsha, 1986

Kume Kunitake. *Tokumei zenken taishi Bei Ō kairan jikki*. Vol. 3. Tokyo: Dajōkan Kirokugakari, 1878.

Kunaichō, eds. *Meiji tennōki*. Vol. 2. Tokyo: Yoshikawa Kobunkan, 1969.

Kusumoto Juichi. *Nagasaki seitetsujo*. Tokyo: Chūkō Shinsho, 1992.

Kuwabara Masa. 'Kamaishi kōzan keikyō hōkoku.' *Kōgaku sōshi*. No.10 (1882): 479–89.

——'Kamaishi kōzan keikyō.' *Kōgaku sōshi*. No.11 (1882): 533–43.

——'Kamaishi kōzan keikyō.' *Kōgaku sōshi*. No.15 (1882): 6–19.

Maeda Masana, *Kōgyō Iken*. Edited by Meiji Bunken Shiryō Kankōkai. *Meiji zenki zaisei keizai shiryō shūsei*. Vol. 18. Tokyo: Meiji Bunken Shiryō Kankōkai, 1962.

Matsukata Masayoshi, *Zaisei kanki gairyaku*, Tokyo: 1880. Edited by Meiji Bunken Shiryō Kankōkai, *Meiji zenki zaisei keizai shiryō shūsei*. Vol. 1. Tokyo: Meiji Bunken Shiryō Kankōkai, 1962.

Matsumura Harusuke. *Kaika senjimon*. Tokyo: Bunkeidō, 1873.

Meiji Bunkō Shiryō Iinkai, ed., *Meiji zenki sangyō hatatsushi shiryō*. Tokyo: Meiji Bunkō Shiryō Iinkai, 1971.

Meiji Zenki Bunken Shiryō Kankōkai, *Meiji zenki sangyō hattatsushi shiryō*, vol. 67:1–4, Tokyo: Meiji Bunken Shiryō Kankōkai, 1970.

Minbushō. *Yōsan shikenhō gofukoku*. Tokyo:1870.

Miyoshi Nobuhiro. *Meiji no enjinia kyōiku: Nihon to Igirisu no chigai*. Tokyo: Chūōkōronsha, 1983.

Mori Kahei. *Iwate-ken no rekishi*. Tokyo: Yamagawa Shuppansha, 1973.

Mori Kahei and Itabashi Gen, eds. *Kindai tetsu sangyō no seiritsu: Kamaishi seitetsujo zenshi*. Kamaishi: Fuji Seitetsu Kabushikigaisha, 1957.

Mori Kōichi. *Nihon kodai bunka no tankyū: tetsu*. Tokyo: Shakai Shisōsha, 1974.

Murakami Shigeyoshi, ed. *Seibun kundoku kindai shōchokushū*. Tokyo: Shin Jinbutsu Ōraisha, 1983.

Muramatsu Teijirō. 'Basuchan–gunkan to kinu no ito.' *Oyatoi gaikokujin*. Vol. 15. *kenchiku–doboku*. Tokyo: Kashimada Shuppankai, 1978.

——*Nihon kindai kenchiku gijutsushi*. Tokyo: Shōkokusha, 1976.

Nagai Yasuoki. *Seishika hikkei*. 3 vols, Tokyo: Yūrindō, 1884.

Naimushō. *Naimushō dai ikkai nenpō*, Tokyo: Naimushō, 1876.

——*Naimushō hotatsu zensho*. Tokyo: Naimushō, 1874–9.

Nakagawa Kōichi. *Sangyō iseki o aruku: kita kantō no sangyō kōkogaku*. Tokyo: Sangyō Gijutsu Senta, 1978.

Nakamura Masanori, Ishii Kanji, and Kasuga Yutaka eds. *Nihon kindai shisō taikei.* Vol. 8, *keizai kōsō.* Tokyo: Iwanami Shoten, 1988.

Nihon Chōki Kyōkai. *Nihon chōki tōkei sōran.* Vol. 2. Tokyo: Nihon Tōkei Kyōkai, 1987.

Nihon Kagakushi Gakkai. *Nihon kagaku gijutsushi taikei.* Tokyo: Daiichi Hōki Shuppan, 1964–69.

Nihon Shiseki Kyōkai, *Ōkubo Toshimichi monjo.* 9 vols. Tokyo: Nihon Shiseki Kyōkai, 1928.

——*Ōkuma Shigenobu kankei monjo.* 6 vols. Tokyo: Tokyo Daigaku Shuppankai, 1935 (1983–85).

Nihon Tekkōshi Hensankai, eds. *Nihon tekkōshi Meiji-hen.* Tokyo: Gogetsu Shobō, 1982.

Nihon Tōkei Kyōkai, eds. *Nihon chōki tōkei sōran.* Tokyo: Nihon Tōkei Kyōkai, 1987.

Nishibori Akira. *Nihon kindaika to Furansu no kōgyō gijutsu: Yokosuka seitetsujo, Yokohama seitetsujo, Ikuno kōzan, Tomioka seishijo.* Tokyo: Surugadai Shuppansha, 1986.

Nishikawa Takeomi. *Bakumatsu Meiji kokusai shijō to Nihon: kiito bōeki to Yokohama.* Tokyo: Yōhikaku Shuppan, 1997.

Noro Kageyoshi. 'Honkoku seitetsugyō no kako oyobi shōrai.' *Tetsu to hagane.* 1 (1916): 1–16.

Nōshōmushō Kōzankyoku. *Kamaishi kōzan chōsa hōkoku.* Tokyo: Nōshōmushō Kōzan Kyoku, 1893.

Nōshōmushō Nōmukyoku. *Daiichiji zenkoku seishikōjo chōsahyō.* Tokyo: Nōshōmushō Nōmukyoku, 1895.

——*Daini seishikōjo chōsahyō.* Tokyo: Nōshōmushō Nōmukyoku, 1898.

Obana Fuyukichi. 'Seitetsujo kensetsuron.' In *Itō Hirobumi hisho ruiten: jitsugyō, kōgyō shiryō.* Tokyo: Hisho Ruiten Iinkai, 1936.

Odaka Atsutada. 'Seishi no hōkoku.' *Ryūmon zasshi.* 60 (15 May 1893): 1–15.

Oe Shinobu. *Nihon no sangyō kakumei.* Tokyo: Iwanami Shoten, 1968.

Ōhashi Shūji, ed. *Bakumatsu Meiji seitetsuron.* Tokyo: Agune, 1991.

Okada Kokichi. *Tatara kara kindai seitetsu e.* Tokyo: Heibonsha, 1990.

Okamoto Yukio and Imatsu Kenji, eds. *Meiji zenki kan'ei kōjō enkaku.* Kyoto: Tōyō Bunkasha, 1983.

Ōkubo Toshiaki. *Meirokusha kō.* Tokyo: Rittaisha, 1976.

Ōkurashō. *Kōgyō iken.* 31 vols. Tokyo: Ōkurashō, 1884.

——*Kōbushō enkaku hōkoku,* Tokyo: Ōkurashō, 1889.

——*Tekkō.* Tokyo: Ōkurashō, 1892.

Ōkurashō, ed. *Meiji zaiseishi.* Vol. 12. Tokyo: Ōkurashōnai Meiji Zaiseishi Hensankai Hensan, 1946.

——*Meiji zenki zaisei keizai shiryō shūsei.* Vol. 17. Tokyo: Hara Shobō, (1931) 1978.

Ōshima, Shinzō. *Ōshima Takatō kōjitsu.* Hyōgō: Ōshima Shinzō, 1937.

Ōtsuka Ryōtarō. *Sanshi.* 2 vols. Tokyo: Fusōen, 1900.

Ōtsuka Takematsu, ed. *Iwakura Tomomi kankei monjo.* Vol. 8. Tokyo: Nihon Shiseki Kyōkai, 1935.

Ōuchi Hyoe, Tsuchiya Takao hen. *Meiji zenki zaisei keizai shiryō shūsei.* Vols. 1, 17, 18. Tokyo: Hara Shobō, 1978–79.

'Rikukaigun kōbu sanshō gōgi shi ichidai seitetsujo o sōken sen to su jōte gaikokujin o yatoire,' *Dajō ruiten,* (gaikoku kokusai), 2: 72, no. 36, (4 May 1875).

Rinji Seitetsu Jigyō Chōsa Iinkai. *Kamaishi oyobi Sennin tetsuzan junshi hōkoku.* Tokyo: Riuji Seitetsu Chosa Iinkai, 1893.

'Ryūshitsu kōmyaku ni atarazu.' *Jiji shinbun.* (10 April 1882) n.p.

Saigusa Hiroto. *Gendai nihon bunmeishi.* Vol. 14. *Gijutsushi.* Tokyo: Toyo Keizai Shinpōsha, 1940.

——'Shoki Kamaishi seitetsushi kenkyū oboegaki.' *Kagakushi kenkyū.* Vol 1. (December 1941): 118–31.
——'Shoki Kamaishi seitetsushi kenkyū oboegaki.' *Kagakushi kenkyū.* Vol 2. (May 1942): 79–92.
——'Shoki Kamaishi seitetsushi kenkyū oboegaki.' *Kagakushi kenkyū.* Vol 3. (November 1942): 108–19.
——*Saigusa Hiroto chōsakushu.* Vols. 9 & 11. Tokyo: Chūōkōronsha, 1972.
Saigusa Hiroto and Iida Ken'ichi, eds. *Nihon kindai seitetsu gijutsu hattatsushi: Yawata seitetsujo no kakuritsu katei.* Tokyo: Tōyō Keizai Shinpōsha, 1957.
Sano Akira. *Dai Nihon sanshi.* 2 vols. Tokyo: Dai Nihon Sanshi Hensan Jimusho, 1898.
Sano Tsunetami. *Ōkoku hakurankai hōkokusho: kōgyō denpa hōkokusho.* Tokyo: Ōkoku Hakurankai Jimukyoku, 1875.
——*Ōkoku hakurankai hōkokusho: sangyōbu.* 2 vols. Tokyo: Ōkoku Hakurankai Jimukyoku, 1875.
Sayatori Motohiko. 'Kōko sōdan.' *Tokyo nichinichi shinbun.* No. 1769 (23 October 1877): 982.
Shashi Hensan Iinkai. *Nichibō 75-nen Shi.* Osaka: Nichibō Kabushiki Kaisha, 1966.
Shibusawa Eiichi. 'Yo wa ika ni shite Tomioka seishijō sekkei kantoku no nin ni atarishia.' *Dai Nihon sanshikaihō,* 200 (1 January 1909): 14.
Shiimada Tadashi. *Za yatoi: oyatoi gaikokujin no sōgōteki kenkyū.* Kyoto: Shinbunkaku Shuppan Co. Ltd., 1987.
Shimokawa Yoshio. *Nihon tekkō gijutsushi.* Tokyo: Agune, 1989.
Shin Nihon Seitetsu Kabushikigaisha. *Honō to tomo ni.* 3 vols. Tokyo: Shin Nihon Seitetsu Kabushikigaisha, 1981.
Soeda Juichi. 'Seitetsusho no setsuritsu wa kyōmu nari.' In Ōkurashō. *Tekkō.* Tokyo: Ōkurashō, 1892.
Sugaya Jūhei. *Nihon tekkōgyō-ron.* Tokyo: Dōbunkan, 1957.
Sugiura Yuzuru Iinkai. *Sugiura Yuzuru zenshū.* 5 vols. Tokyo: Sugiura Yuzuru Iinkai, 1979.
Suzuki Jun. *Meiji no kikai kōgyō: sono seisei to hattatsu.* Tokyo: Minerva, 1996.
Suzuki Jun, ed., *Kōbushō to sono jidai,* Tokyo: Yamakawa Shuppansha, 2002.
Tachi Saburō. *Kiito seihō shin'an.* Tokyo: Yoshidaya, 1874.
Taigakai hen. *Naimushō shi.* Tokyo: Hara Shobō, 1980–1.
'Taiwan shusō.' *Tokyo nichi nichi shinbun.* No. 770 (14 August 1874) n.p.
Takamatsu Tōru. 'Kamaishi Tanaka seitetsujo mokitankoro no tekkan neppurō.' *Gijutsu to Bunmei.* 10 (June 1990): 47–67.
Tamura, Eitaro. *Nihon no gijutsusha.* Tokyo: Koa Shobō, 1943.
Tokutomi Sohō. *Kōshaku Matsukata Masayoshi den.* Vol. 1. Tokyo: Kōshaku Matsukata Masayoshi Denki Hakkōjo, 1935.
Tokyo Daigaku Shakai Kagaku Kenkyōjo, ed. *20 seiki shisutemu: 4 kaihatsushugi.* Tokyo: Tokyo Daigaku Shuppankai, 1998.
Tokyo Keizai Daigaku. *Ōkurasho gyōgakkō ni okeru meishi enzetsusho.* preliminary edition. Tokyo: Tokyo Keizai Daigaku, 1985.
'Tomioka seishijo no seihin ke kaite ki ni yumei.' *Tokyo nichi nichi shinbun.* (6 June 1873) n.p.
Tomioka Seishijōshi Hensan Iinkai, ed. *Tomioka seishijōshi,* 2 vols. Tomioka: Tomioka-shi Kyōiku Iinkai, 1977.
Tomita Masafumi and Tsuchibayashi Shunichi, eds. *Fukuzawa Yukichi zenshū.* Vols. 4, 5, 7, 8, 10, 12, 14. Tokyo: Iwanami Shoten, 1959.
Tsuchiya Takeo. *Zoku Nihon keizaishi gaiyō.* Tokyo: Iwanami, 1941.

Tsuchiya Takao, ed. *Shibusawa Eiichi denki shiryō*. Vols 2 & 3. Tokyo: Ryūmonsha, 1956.
Tsuji Zennosuke. *Nihon bunkashi*. Vol. 7. Tokyo: Shunjōsha, 1950.
Tsukahara Ryōgo, ed. *Rankō ō*. Tokyo: Takahashi Hiratarō, 1909.
Tsūshō Sangyōsho hen. *Shōkō seisakushi*. Vol. 17. Tokyo: Shōkō Seisakushi Kankōkai, 1962.
Ueda Akikizo. 'Kōzan gichō Gottofurei.' *Shōwa 60 nendō zenkoku chika shigen kankei gakukyōkai gōdō shūki taikai bunka kenkyūkai shiryō*. G2 (October, 1986): 5–8.
Wada Ei. *Tomioka nikki*. Kamijō Hiroyuki, ed. Tokyo: Sojusha, 1976.
Waseda Daigaku Shakai Kagaku Kenkyūjo, ed. *Ōkuma bunsho*. Vol. 2. Tokyo: Waseda Daigaku Shakai Kagaku Kenkyūjo, 1959.
Watanabe Wataru. *Kamaishi kōzan keikyō hōkokura*. Vol. 16, Meiji-hen, mae, 1. Tokyo: Nihon Kōgyōshi Shiryō Kankō Iinkai, 1994.
——*Kōgyō kaisha teisoku*. Vol. 16, Meiji-hen, mae, 2. Tokyo: Nihon Kōgyōshi Shiryō Kankō Iinkai, 1994.
Yokohama Kaikō Shiryōkan, ed. *Yokohama gaikokujin kyoryūchi*. Yokohama: Yokohama Kaikō Shiryōkan, 1998.
Yoshida Mitsukuni. *Oyatoi gaikokujin*. 12 vols. Tokyo: Kajima Kenkyūjo Shuppankai, 1968.
Yoshii Tōru. *Kōgyō yōsetsu*. Tokyo: Yoshii Tōru, 1880. Also available as *Yoshii Tōru, Kōgyō yōsetsu*, in Nihon kōgyō hirāshū. Tokyo: Hakua shobō, 1984.
Yoshimi Shunya. *Hakurankai no seijigaku: manzanshi no kindai*. Tokyo: Chūōkōronsha, 1992.
'Zatsuroku: Yue meiyo kai'in Itō Yajirō kun no itsuiji.' *Nihon kōgyō kaishi*. 46: 547 (November 1930): 1055.

Western-language materials

Abe, Takeshi. 'Technological and Organizational Absorption in the Development of the Modern Japanese Cotton Industry.' Fifth Conference of the Global Economic History Network. Osaka, Japan, 16–18 December 2004. Available at www.lse.ac.uk/collections/economicHistory/GEHN/GEHNPDF/Abe(text)GEHN5.pdf (accessed 19 December 2006).
Abosch, David. *Kato Hiroyuki and the Introduction of German Political Thought in Modern Japan*. Ann Arbor, MI: University Microforms, 1964.
Adas, Michael. *Machines as the Measure of Men: Science, Technology, and Ideologies of Western Dominance*. Ithaca, NY: Cornell University Press, 1989.
Aikens, Melvin C. and Higuchi Takayasu. *The Prehistory of Japan, Studies in Archaeology*. New York: Academic Press, 1982.
Anderson, Kym, ed. *New Silk Roads: East Asia and World Textile Markets*. Cambridge: Cambridge University Press, 1991.
Aoki Eiichi, 'Dawn of Japanese Railways.' *Japan Railway and Transport Review*. (March, 1994): 28–30.
——'Expansion of Railway Network.' *Japan Railway and Transport Review*. (June, 1994): 34–7.
——'Growth of Independent Technology.' *Japan Railway and Transport Review*. (October, 1994): 56–9.
Bartholomew, James R. *The Formation of Science in Japan*. New Haven: Yale University Press, 1989.

Bavier, Ernest de. *La Sériciculture le Commerce des Soies et des Graines et L'Industrie de la Soie au Japon*. Lyon: H. Georg, 1874.

Beasley, William G. *The Modern History of Japan*. New York: Frederick Praeger, 1963.

——*The Meiji Restoration*. Stanford, CA: Stanford University Press, 1972.

——*Japan Encounters the Barbarian: Japanese Travellers in America and Europe*. New Haven, CT: Yale University Press, 1995.

Beauchamp, Edward R. and Akira Iriye, eds. *Foreign Employees in Nineteenth Century Japan*. Boulder, CO: Westview Press, 1990.

Bektas, Yakup. 'Displaying the American Genius: The electromagnetic telegraph in the wide world.' *British Journal for the History of Technology*. 34:121 (June 2001): 199–232.

Bell, S. Peter. *A Biographical Index of British Engineers in the 19th Century*. New York: Garland, 1975.

Beranek William Jr., and Gustav Ranis, eds. *Science Technology and Economic Development*. New York: Praeger, 1978.

Betz, Mathew J., Pat McGowan, and Rolf T. Wigand, eds. *Appropriate Technology: Choice and Development*. Durham, NC: Duke Press Policy Studies, 1984.

Bijker, Wiebe E. and John Law, eds. *Shaping Technology/Building Society: Studies in Sociotechnical Change*. Cambridge: MIT Press, 1992.

Bijker, Wiebe E., Thomas P. Hughes, and Trevor Pinch. *The Social Construction of Technological Systems: New Directions in the Sociology and History of Technology*. Cambridge: MIT Press, 1989.

Birkinbine, John. 'Charcoal as a Fuel for Metallurgical Processes.' *Transactions of the American Institute of Mining Engineers*. 11 (May 1882–Feb. 1883): 78–88.

Blacker, Carmen. *The Japanese Enlightenment: A Study of the Writings of Fukuzawa Yukichi*. Cambridge: Cambridge University Press, 1964.

Borton, Hugh. *Japan's Modern Century*. New York: The Ronald Press, 1955.

Bourdieu, Pierre. *The Logic of Practice*. Richard Nice, trans. Cambridge: Polity Press, 1990.

Braguinsky, Serguey, Atsushi Ohyama, and David C. Rose. 'Cooperative Technology Adoption Under Global Competition: The Case of the Japanese Cotton Spinning Industry.' *George J. Stigler Center for the Study of the Economy and the State Working Papers Series*. 27 July 2002. Available at www.isnie.org/ISNIE02/Papers02/braguinsky. pdf (accessed 15 December 2006).

Braisted, William Reynolds, trans. *Meiroku Zasshi: Journal of the Japanese Enlightenment*. Tokyo: University of Tokyo Press, 1976.

Brock, William H. 'David Forbes.' *New Dictionary of National Biography*. p. 824.

Brown, Sidney Devere. 'Ōkubo Toshimichi: His Political and Economic Policies in Early Meiji Japan,' *Journal of Asian Studies* 21:2 (February 1962): 183–97.

Brown, Sidney Devere and Akiko Hirota. trans. *The Diary of Kido Takayoshi*. 3 vols. Tokyo: University of Tokyo Press, 1986.

Brown, M. Walton. 'Coal-Fields of Japan.' *Transactions of the Federated Institution of Mining Engineers*. 10 (1895–6): 538–43.

Brunton, Richard Henry. 'Public Works.' *The Japan Mail*. (25 September 1875): 558–64.

——'The Japan Lights.' *Proceedings of the Institution of Civil Engineers*. Vol. xlvii, 1876–7.

——*Schoolmaster to an Empire, Richard Henry Brunton in Meiji Japan, 1868–1876*. Edward R. Beauchamp, ed. New York: Greenwood Press, 1991.

——*Building Japan. 1868–1876*. With introduction and notes by Sir Hugh Cortazzi, Sandgate: Japan Library, Ltd., 1991.

Buchanan, Daniel H. *The Development of Capitalistic Enterprise in India*. New York: MacMillan, 1934.

Buckle, Henry Thomas. *History of Civilization in England*. Vol. I. New York: D. Appleton, 1859.

Bureau of Mines. *Mining in Japan, Past and Present*. Tokyo: The Bureau of Mines, The Department of Agriculture and Commerce of Japan, 1909.

Burke-Gaffney, Brian. 'Hashima: The Ghost Island.' *Crossroads*. 4 (Summer, 1996): 33–52.

Bury, J. B. *The Idea of Progress: an Inquiry into its Origin and Growth*. New York: Macmillan, 1932.

Cawley, George. 'Some Remarks on Construction in Brick and Wood and their Relative Suitability for Japan.' *Transactions of the Asiatic Society of Japan*. 6: 2 (1878): 291–317.

Centre for East Asian Cultural Studies, compiler. *Meiji Japan Through Contemporary Sources*. 3 vols. Tokyo: Centre for East Asian Cultural Studies, 1969–72.

Chaiklin, Martha. 'A Miracle of Industry: The Struggle to Produce Sheet Glass in Modernizing Japan.' In Morris F. Low, ed. *Building Modern Japan: Science, Technology, and Medicine in the Meiji Era and Beyond*. New York: Palgrave Macmillan, 2005.

Chamarik, Saneh and Susantha Goonatilake, eds. *Technological Independence: The Asian Experience*. Tokyo: United Nations University Press, 1994.

Chida, Tomohei and Peter N. Davies. *The Japanese Shipping and Shipbuilding Industries: a History of Their Modern Growth*. London: Athlone Press, 1990.

Cho, S., and Runeby, N., eds. *Traditional Thought and Ideological Change: Sweden and Japan in the Age of Industrialization*. Stockholm: Department of Japanese and Korean, University of Stockholm, 1988.

Church, R. A., ed. *The Coal and Iron Industries*, Oxford: Blackwell, 1994.

Clancey, Gregory K. *Foreign Knowledge or Art Nation, Earthquake Nation: Architecture, Seismology, Carpentry, the West, and Japan, 1876–1923*, Ph.D. dissertation, Cambridge: Massachusetts Institute of Technology, 1998.

——*Earthquake Nation: The Cultural Politics of Japanese Seismicity, 1868–1930*. Berkeley: University of California Press, 2006.

Clyde, Paul H. compiler. *United States Policy Toward China: Diplomatic and Public Documents, 1839–1939*. Durham, NC: Duke University Press, 1940.

Cobbing, Andrew. *The Japanese Discovery of Victorian Britain: Early Travel Encounters in the Far West*. Surrey: Japan Library (Curzon Press), 1998.

Coignet, Francisque. *Note sur la richesse minérale du Japon*. In Junkichi Ishikawa, *Nihon kōbutsu shigen ni kansuru oboegaki*. Vol. 7. Tokyo: Haneda Shoten, 1944.

Collins, Joseph Henry. *First Book of Mineralogy*. London and Glasgow: Collins, 1873.

——*Principles of Coal Mining*. New York: Putnam, 1986.

——*Principles of Metal Mining*. New York: Putnam, 1876.

Conroy, Hilary. *The Japanese Seizure of Korea: 1868–1910, A Study of Realism and Idealism in International Relations*. Philadelphia, PA: University of Pennsylvania Press, 1960.

Conroy, Hilary, Sandra T. W. Davis, and Wayne Patterson, eds. *Japan in Transition: Thought and Action in the Meiji Era, 1868–1912*. Rutherford: Fairleigh Dickinson University Press, 1984.

Constant, Edward W. *The Origins of the Turbojet Revolution*. Baltimore, MD: Johns Hopkins University Press, 1980.

'Correspondances,' *Moniteur des Soies*, 17 September 1870, n.p.

Cortazzi, Hugh and Gordon Daniels, eds. *Britain and Japan 1859–1991: Themes and Personalities*. London: Routledge, 1991.

Craig, Teruko, trans., *The Autobiography of Shibusawa Eiichi: From Peasant to Entrepreneur*. Tokyo: University of Tokyo Press, 1994.

Cusumano, Michael A. 'An Enlightenment Dialogue with Fukuzawa Yukichi: Ogawa Tameji's *Kaika Mondō*, 1874–1875.' *Monumenta Nipponica*. 37: 3 (Autumn 1982): 375–401.

Davis, Sandra T. W. 'Ono Azusa and the Political Change of 1881.' *Monumenta Nipponica*. 25: 1/2 (1970): 137–54.

Debin Ma, 'The Modern Silk Road: The Global Raw-Silk Market, 1850–1930,' *The Journal of Economic History*, 56: 2, (June, 1996), p. 341.

Dilke, Charles Wentworth. 'English Influence in Japan.' *Fortnightly Review*. Vol. XX , new series (1 July to 1 December 1876): 424–43.

Dosi, Giovanni, Renato Giannetti, and Pier Angelo Toninelli, eds. *Technology and Enterprise in a Historical Perspective*. Oxford: Clarendon Press, 1992.

Duncan, P. Martin. 'Presidential Address.' *Proceedings of the Geological Society of London*. vol. 33, supplement (1877): 41–7.

Duus, Peter. *The Abacus and the Sword: The Japanese Penetration of Korea, 1895–1901*, Berkeley, CA: University of California Press, 1995.

Dyer, Henry, *The Evolution of Industry*. New York: Macmillan, 1895.

——*Dai Nippon: The Britain of the East*, London: Blackie, 1904.

Ehlers, J. H. *Raw Materials Entering into the Japanese Iron and Steel Industry*. United States Department of Commerce, Trade Bulletin No. 573, 1928.

Ehrlich, Eva. *Japan, a Case of Catching-up*. Budapest, Akadémiai Kaidó, 1984.

Ericson, Steven. *Sound of the Whistle: Railroads and the State in Meiji Japan*. Cambridge: Harvard University Press, 1996.

——'Importing Locomotives in Meiji Japan' International Business and Technology Transfer in the Railroad Industry.' *Osiris*. 2nd series, 13 (1998–99): 129–54.

Ewen, Stuart. *All Consuming Images: The Politics of Style in Contemporary Culture*. New York: Basic Books, 1988.

'Export Trade.' *North China Herald*. (5 February 1874): 123.

'Export Trade.' *North China Herald*. (22 August 1874): 205.

'Export Trade.' *North China Herald*. (25 May 1880): 467.

Federico, Giovanni. *An Economic History of the Silk Industry, 1830–1930*. Cambridge University Press, 1997.

Feenberg, Andrew and Alastair Hannay. *Technology and the Politics of Knowledge*. Bloomington, IN: Indiana University Press, 1995.

Finn, Dallas. *Meiji Revisited: Sites of Victorian Japan*. New York: Weatherhill, 1995.

'The Follies of Imitation.' *North China Herald*. (28 March 1878): 32.

Forbes, David. 'Report on the Progress of the Iron and Steel Industries in Foreign Countries.' *Journal of the Iron and Steel Institute*. (1875, part 1). London: The Iron and Steel Institute.

——'Report on the Progress of the Iron and Steel Industries in Foreign Countries.' *Journal of the Iron and Steel Institute*. (1875, part 2). London: The Iron and Steel Institute.

Francks, Penelope. *Technology and Agricultural Development in Pre-War Japan*. New Haven, CT: Yale University Press, 1984.

——*Japanese Economic Development: Theory and Practice*. London: Routledge, 1992.

Fransman, Martin, ed. *Machinery and Economic Development*. London: MacMillan, 1986.

Fujita, Fumiko. *American Pioneers and the Japanese Frontier: American Experts in Nineteenth-Century Japan*. Westport, CT: Greenwood Press, 1994.

Fukasaku Yukiko. *Technology and Industrial Development in Pre-war Japan: Mitsubishi Nagasaki Shipyard, 1884–1934*. London: Routledge, 1992.

Fukuzawa Yukichi. *An Encouragement of Learning*. David A. Dilworth and Umeyo Hirano, trans., Tokyo: Sophia University Press, 1969.

——*An Outline of a Theory of Civilization*. David A. Dilworth and G. Cameron Hurst, trans., Tokyo: Sophia University Press, 1973.

Galtung, Johan. *Development, Environment, and Technology: Towards a Technology of Self-reliance*. New York: United Nations Publications, 1979.

Gerschenkron, Alexander. 'Economic Backwardness in Historical Perspective' in *Economic Backwardness in Historical Perspective, A Book of Essays*. Cambridge: Belknap Press, 1962.

Godfrey, J. G. H. 'Notes on the Geology of Japan.' *The Quarterly Journal of the Geological Society of London*. 34 (1878): 542–55.

Gooday, Graeme J. N. and Morris F. Low. 'Technology Transfer and Cultural Exchange: Western Scientists and Engineers Encounter Late Tokugawa and Meiji Japan.' *Osiris*. 2nd series, 13 (1998–99): 99–128.

Gordon, Andrew. *The Evolution of Labor Relations in Japan: Heavy Industry, 1853–1955*. Cambridge, MA: Harvard University Press, 1985.

Gotō Akira and Hiroyuki Odagiri, eds. *Innovation in Japan*. Oxford: Clarendon Press, 1997.

Goulet, Denis. *The Uncertain Promise: Value Conflicts in Technology Transfer*. New York: New Horizons Press, 1989.

Gowland, William. 'Metals and Metalworking in Old Japan.' *Transactions and Proceedings of the Japan Society of London*. 13 (1915): 21–101.

'The Grand Tour.' *Japan Weekly Mail*. (15 July 1871): 389–91.

Greenwood, William H. *A Manual of Metallurgy*. Vol. 1. London: William, 1874.

Gruber, William H. and Donald G. Marqui, eds. *Factors in the Transfer of Technology*. Cambridge, MA: MIT Press, 1969.

Gueneau Louis. *Lyon et le Commerce de la Soie*. Lyon: L. Bascou, 1923.

Guizot, Francois Pierre Guillaume. *History of Civilization in Europe*. George Wells Knight, ed., New York: D. Appleton, 1896.

Habermas, Jürgen, *The Philosophical Discourse of Modernity: Twelve Lectures*. Frederick Lawrence, trans., Cambridge, MA: MIT Press, 1987.

Hall, Ivan P. *Mori Arinori*. Cambridge, MA: Harvard University Press, 1973.

Hall, John Whitney. *Japan from Prehistory to Modern Times*. New York: Delacorte Press, 1968.

Hanley, Susan B. *Everyday Things in Premodern Japan: The Hidden Legacy of Material Culture*. Berkeley, CA: University of California Press, 1997.

Hao, Yen-p'ing and Erh-min Wang. 'Changing Chinese Views of Western Relations, 1840–1895.' In *Cambridge History of China*. Vol. 11, part 2. Dennis Twitchett and John K. Fairbanks, eds. Cambridge: Cambridge University Press, 1980.

Harada Katsumasa. 'The Technical Progress of Railways in Japan in Relation to the Policies of the Japanese Government.' In *Papers of the History of Industry and Technology in Japan*. Vol. 2. Erich Pauer, ed. Marburg: Förderverein Marburger Japan-Reihe, 1995.

Hård, Mikael and Andrew Jamison. *The Intellectual Appropriation of Technology: Discourses on Modernity, 1900–1939*. Cambridge, MA: MIT Press, 1998.

Harris, J. R. *The British Iron Industry, 1700–1850*. London: Macmillan Education, 1988.

Hashino Tomoko and Osamu Saito. 'Tradition and Interaction: Research Trends in Modern Japanese Industrial History.' *Australian Economic History Review*. 44:3 (November 2004): 241–58.

Hastings, Sally. 'The Empress' New Clothes and Japanese Women, 1868–1912.' *The Historian*. 55: 4 (Summer 1993): 677–92.

Hattori, Susumu. 'The Iron and Steel Industries of Japan. A Review of its Present Condition with an Outline of its Historical Development.' *Proceedings of the World Engineering Conference*. 33: 1(1931): 43–78.

Hayashi, Takeshi. *The Japanese Experience in Technology: From Transfer to Self-Reliance.* Tokyo: United Nations University Press, 1990.

Hayes, Carlton J. H. *A Generation of Materialism, 1871–1900.* New York: Harper & Row, 1941.

Hess, David J. *Science and Technology in a Multicultural World: The Cultural Politics of Facts and Artifacts.* New York: Columbia University Press, 1995.

Higonnet, Patrice, David S. Landes, and Henry Rosovsky, eds. *Favorites of Fortune, Technology, Growth, and Economic Development since the Industrial Revolution.* Cambridge, MA: Harvard University Press, 1991.

Holland, Max. *When the Machine Stopped: A Cautionary Tale from Industrial America.* Boston, MA: Harvard Business School Press, 1989.

Hopper, David H. *Technology, Theology and the Idea of Progress.* Louisville, KY: Westminster/John Knox Press, 1991.

Horie Yasuzō. 'Government Industries in the Early Years of the Meiji Era.' *Kyoto University Economic Review.* XIV (1939): 67–87.

——'Business Pioneers of Modern Japan.' *Kyoto University Economic Review.* 30:2 (October 1960): 1–16.

Hunter, Janet. 'Regimes of Technology Transfer in Japan's Cotton Industry 1860s–1890s.' Paper presented at the ninth conference of the Global Economic History Network, Kaohsiung, Taiwan, 9–11 May 2006. Available at www.wtuc.edu.tw/intaffairs/international%20cooperation/Papers/Regimes%20of%20Technology%20Transfer%20in%20Japan.doc (accessed 15 December 2006).

Huber, Charles J. *The Raw Silk Industry of Japan.* New York: The Silk Association of America, 1929.

Iheukwu, Anthony C. *Technology Dependence and Transfer.* M. A. Thesis. Columbus, OH: The Ohio State University: 1983.

Ike Nobutaka. *The Beginnings of Political Democracy in Japan.* Baltimore, MD: The Johns Hopkins Press, 1950.

Imperial Bureau of Mines. *Mining Industry in Japan, Prepared for the World's Panama Pacific Exposition, San Francisco, Cal., 1915.* Tokyo: Imperial Bureau of Mines, 1915.

Inkster, Ian. *Japan as a Development Model? Relative Backwardness and Technological Transfer.* Bochum, Germany: Studienverlag Brockmeyer, 1980.

——*Science and Technology in History: An Industrial Approach to Industrial Development.* New Brunswick, NJ: Rutgers University Press, 1991.

International Exhibition of 1876: Official Catalogue of the Japanese Section, and Descriptive notes on the Industry and Agriculture of Japan. Philadelphia, PA: Japanese Commission, 1876.

Inukai Ichirō. 'The Kōgyō Iken: Japan's Ten Year Plan, 1884.' *KSU Economics and Business Review.* 6 (May 1979): 1–100.

Inwood, Kris E. *The Canadian Charcoal Iron Industry 1870–1914.* New York: Garland Publishing, 1986.

Islam, Rizwanul, ed. *Transfer, Adoption and Diffusion of Technology for Small and Cottage Industries.* Geneva, Switzerland: International Labour Organization, 1992.

Itō Takashi and George Akita. 'The Yamagata-Tokutomi Correspondence: Press and Politics in Meiji-Taisho Japan.' *Monumenta Nipponica.* 36: 4 (Winter, 1981): 391–423.

Iwata, Masakazu. *Okubo Toshimichi, the Bismarck of Japan.* Berkeley: University of California Press, 1964.

Japan Biography Research Department. *Japan Biographical Encyclopedia and Who's Who.* Tokyo: The Rengo Press, 1958.

'Japanese Dress for the Japanese.' *The Japan Times.* (6 April 1878): 292.

'Japanese News – Court, Political, Official.' *Japan Weekly Mail.* (11 September 1880): 1180.

'Japanese Vanity.' *Japan Weekly Mail.* (11 February 1871): 67–8.

Jansen, Marius B., ed. *Changing Japanese Attitudes Toward Modernization.* Rutland, VT: Charles E. Tuttle, 1982.

——*The Cambridge History of Japan.* Vol. 5. Cambridge: Cambridge University Press, 1989.

Jeremy, David J., ed. *International Technology Transfer, Europe, Japan, and the USA, 1700–1914.* London: Edward Elgar, 1991.

Johnson, Chalmers. *MITI and the Japanese Miracle: The Growth of Industrial Policy, 1925–1975.* Stanford, CA: Stanford University Press, 1982.

Johnson, William A. *The Steel Industry of India.* Cambridge: Harvard University Press, 1966.

Jones, Hazel J. 'The Formulation of the Meiji Government Policy Toward the Employment of Foreigners.' *Monumenta Nipponica.* 23: 1–2 (1968): 9–30.

——*Live Machines: Hired Foreigners and Meiji Japan.* Vancouver: University of British Columbia Press, 1980.

——'Live Machines Revisited.' In Edward R. Beauchamp and Akira Iriye, eds. *Foreign Employees in Nineteenth-Century Japan.* Boulder, CO: Westview Press, 1990.

Jorden, Paul. 'Mineral Resources of Japan.' *Transactions of the Federated Institution of Mining Engineers.* 16 (1898–99): 530–2.

Katō Kōzaburō. 'Yamanobe Takeo and the Modern Cotton Spinning Industry in the Meiji Era.' In Erich Pauer, ed. *Papers of the History of Industry and Technology in Japan.* Vol. II. Marburg: Förderverein Marburger Japan-Reihe, 1995.

'Kamaishi.' *The Japan Weekly Mail.* (28 August 1875): 740–2.

Kelly, Allen C. and Jeffrey G. Williamson. *Lessons From Japanese Development.* Chicago, IL: University of Chicago Press, 1974.

Kemp, Tom. *Industrialization in the non-Western World.* New York: Longman, 1983.

——*Historical Patterns of Industrialization.* Second edition. London: Longman, 1993.

Kennichi Ohno, *The Economic Development of Japan: The Path Traveled by Japan as a Developing Country.* Tokyo: GRIPS Development Forum, 2006.

Kenwood, A. G. and A. L. Lougheed. *Technological Diffusion and Industrialisation Before 1914.* London: Croom Helm, 1982.

Kim, Soon-Ja. *Historical Development of Japanese Secondary Technical Education 1870–1935.* Ph. D. dissertation, Pittsburgh: University of Pittsburgh, 1978.

Kinmonth, Earl H. 'Fukuzawa Reconsidered: *Gakumon no susume* and Its Audience.' *Journal of Asian Studies.* 37: 4 (August 1978): 677–96.

——*The Self-Made Man in Meiji Japanese Thought: From Samurai to Salary Man.* Berkeley, CA: University of California Press, 1981.

Kiyokawa Yukihiko. 'Transplantation of the European Factory System and Adaptations in Japan: The Experience of the Tomioka Model Filature.' *Hitotsubashi Journal of Economics.* 28 (1987): 27–39.

——'The Transformation of Young Rural Women into Disciplined Labor Under Competition-Oriented Management: The Experience of the Silk-Reeling Industry in Japan.' *Hitotsubashi Journal of Economics.* 32 (1991): 49–69.

Klein, Lawrence and Kazushi Ohkawa. *Economic Growth: The Japanese Experience Since the Meiji Era.* Homewood, IL: Richard D. Irwing, 1968.

Kline, Ronald and Trevor Pinch. 'Users as Agents of Technological Change: The Social Construction of the Automobile in the Rural United States.' *Technology and Culture.* 37: 4 (Oct. 1996): 763–96.

Kornicki, Peter F. 'Public Display and Changing Values: Early Meiji Exhibitions and Their Precursors.' *Monumenta Nipponica*. 49: 2 (Summer 1994): 167–96.

Kosaka, Masaaki. *Japanese Thought in the Meiji Era*. Translated and adapted by David Abosch. Tokyo: Pan-Pacific Press, 1958.

Kozo Yamamura. 'Success Illgotten? The Role of Meiji Militarism in Japan's Technological Progress.' *Journal of Economic History*. 37: 1 (March 1977).

Krogh, Thomas. *Technology and Rationality*. Aldershot: Ashgate, 1998.

Kuhn, Thomas S. *The Structure of Scientific Revolutions*. Chicago, IL: University of Chicago Press, 1970.

Kuroda Taizo. 'On Blast Furnace Coke and Fire Bricks for the Iron and Steel Industries in Japan.' *Proceedings of the World Engineering Conference*. 33: 1 (1931): 133–62.

Lakatos, Imre and Alan Musgrave. *Criticism and the Growth of Knowledge*. Cambridge: Cambridge University Press, 1970.

Landes, David S. *The Unbound Prometheus: Technological Change and Industrial Development in Western Europe from 1750 to the Present*. Cambridge, MA: Harvard University Press, 1969.

Latour, Bruno. *Science in Action: How to Follow Scientists and Engineers Through Society*. Cambridge, MA: Harvard University Press, 1987.

Latour, Bruno and Steve Woolgar. *Laboratory Life: The Social Construction of Scientific Facts*. Beverly Hills, CA: Sage, 1979.

Layton, Edwin T. *The Revolt of the Engineers: Social Responsibility and the American Engineering Profession*. Baltimore, MD: Johns Hopkins University Press, 1986.

Lebra, Joyce Chapman. 'Ōkuma Shigenobu and the 1881 Political Crisis.' *Journal of Asian Studies*. 18: 4 (August 1959): 475–87.

——*Ōkuma Shigenobu: Statesman of Meiji Japan*. Canberra: Australian National University Press, 1973.

Liu, Ta-chun. *The Silk Industry of China*. Shanghai: Kelly and Walsh, 1941.

Lockwood, William W. ed. *The State and Economic Enterprise in Japan*. Princeton, NJ: Princeton University Press, 1965.

——*The Economic Development of Japan: Growth and Structural Change*. Princeton, NJ: Princeton University Press, 1968.

Lubar, Steven and W. David Kingery, eds. *History from Things: Essays on Material Culture*. Washington, DC: Smithsonian Institution Press, 1993.

Lyman, Benjamin Smith. *Geological Survey of Japan: Report on the Second Year's Progress of the Survey of the Oil Lands of Japan*. Tokyo: Public Works Department, 1878.

Lynn, Leonard H. *How Japan Innovates: A Comparison with the U. S. In the Case of Oxygen Steelmaking*. Boulder, CO: Westview Press, 1982.

McCallion, Stephen. *Silk Reeling in Japan: The Limits to Change*. Ph. D. dissertation. Columbus, OH: The Ohio State University, 1983.

Mackay, Hughie and Gareth Gillespie. 'Extending the Social Shaping of Technology Approach: Ideology and Appropriation.' *Social Studies of Science*. 22: 4 (November 1992): 685–716.

MacKenzie, Donald and Judy Wajcman, eds. *The Social Shaping of Technology: How the Refrigerator got its Hum*. Philadelphia, PA: Open Press, 1985.

McMaster, John. 'The Takashima Mine: British Capital and Japanese Industrialization.' *Business History Review*. 37: 3 (Autumn, 1963): 217–39.

Manning, George K., ed. *Technology Transfer: Successes and Failures*. San Francisco, CA: San Francisco Press, 1974.

Maquet, Jacques. *Civilizations of Black Africa*. New York: Oxford University Press, 1972.

Marx, Leo and Bruce Mazlish, eds. *Progress: Fact or Illusion?*. Ann Arbor, MI: The University of Michigan Press, 1996.

Mason, Mark. *American Multinationals and Japan: The Political Economy of Japanese Capital Controls, 1899–1980*. Cambridge, MA: Harvard University Press, 1992.

Mass, William and Andrew Robertson, 'From Textiles to Automobiles: Mechanical and Organizational Innovation in the Toyoda Enterprises, 1895–1933.' *Business and Economic History*. 25: 2 (Winter, 1996): 1–37.

Mathiopoulos, Margarita. *History and Progress: In Search of the European and American Mind*. New York: Praeger, 1989.

Mayo, Marlene. 'The Western Education of Kume Kunitake, 1871–6.' *Monumenta Nipponica*. 28: 1 (1973): 3–67.

'Memoirs,' *Minutes and Proceedings of the Institution of Civil Engineers*. XLIX (1877): 273.

'Memoirs of Deceased Members: David Forbes.' *Proceedings of the Institution of Civil Engineers*. 49: 3 (1877): 270–5.

Menzies, W. J. 'The Ashio Copper Mines and Smelting Works of Japan, (abstract).' *Transactions of the Federated Institution of Mining Engineers*. 3 (1891–92): 1061.

Minami, Ryōshin, Ralph Thompson, and David Merriman, trans. *The Economic Development of Japan: a Quantitative Study*. Basingstoke: Macmillan, 1986.

Minami, Ryōshin, Kwan S. Kim, Fumio Makino, and Joung–hae Seo, eds. *Acquiring, Adapting and Developing Technologies: Lessons from the Japanese Experience*. New York: St. Martin's Press, 1995.

'Mining Productions.' *The Japan Weekly Mail*. (23 November 1878): 1271.

Mish, Frederick C., ed., *et al*. *Webster's Ninth New Collegiate Dictionary*. Springfield: Merriam-Webster, 1987.

Mitcham, Carl. *Thinking Through Technology: The Path Between Engineering and Philosophy*. Chicago, IL: University of Chicago Press, 1994.

——ed. *Research in Philosophy and Technology*. Vol. 16, 'Technology and Social Action.' Greenwich, CT: JAI Press, 1997.

Mitcham, Carl, and Robert Mackey. *Philosophy and Technology: Readings in the Philosophical Problems of Technology*. London: Free Press, 1983.

'Modern Japan.' *Japan Weekly Mail*. (28 May 1878) n.p.

'Modern Japan.' *The North China Herald*. (8 June 1878): 588–9.

Mondy, Edmund F. 'Japanese Metallurgical Processes–No. I.' *Engineering*. (19 March 1880): 217–8.

Monroe, Henry S. 'The Mineral Wealth of Japan.' *Transactions of the American Institute of Mining Engineers*. 5 (1876–7): 236–302.

Morris, John. 'Memoirs of the Late Mr. David Forbes.' *Journal of the Iron and Steel Institute*. (1876, part II): 523.

Morris-Suzuki, Tessa. *The Technological Transformation of Japan: From the Seventeenth to the Twenty-first Century*. Cambridge: Cambridge University Press, 1994.

Motoyama Yukihiko. *Proliferating Talent: Essays on Politics, Thought, and Education in the Meiji Era*. Honolulu, HI: University of Hawaii Press, 1997.

Mumford, Lewis. *Technics and Civilization*. New York: Harcourt, Brace, 1934.

Muramatsu, Teijirō. *Industrial Technology in Japan: A Historical Overview*. Tokyo: Hitachi, 1968.

Najita Tetsuo. *Japan: The Intellectual Foundations of Modern Japanese Politics*. Chicago, IL: University of Chicago Press, 1974.

Najita Tetsuo and Irwin Scheiner, eds. *Japanese Thought in the Tokugawa Period, 1600–1868: Methods and Metaphors*. Chicago, IL: University of Chicago Press, 1978.

Nakamura Naofumi. 'Meiji-era Industrialization and Provincial Vitality: The Significance of the First Enterprise Boom of the 1880s.' *Social Science Japan Journal*. 3: 2 (October 2000): 187–205.

Nature. 6 (25 July 1872): 250.

Netto, Curt. *Memoirs of the Science Department, University of Tokio, Japan*. Vol. 2. 'On Mining and Mines in Japan.' Tokyo: University of Tokio, 1879.

Nisbet, Robert. *History and the Idea of Progress*. New York: Basic Books, 1980.

Nish, Ian. *A Short History of Japan*. London: Praeger, 1968.

Nish, Ian, ed. *Britain & Japan: Biographical Portraits*. Folkestone: Japan Library, 1994.

——*Britain & Japan: Biographical Portraits*. Vol. 2. Folkestone: Japan Library, 1997.

——*The Iwakura Mission in America and Europe: A New Assessment*. Richmond: Japan Library (Curzon Press Ltd.) 1998.

Nishio Keijiro. 'The Mining Industries of Japan.' *Transactions of the American Institute of Mining Engineers*. 43 (1913): 54–98.

Nivison, David S. and Arthur F. Wright, eds. *Confucianism in Action*. Stanford, CA: Stanford University Press, 1959.

Noble, David F. *Forces of Production*. New York: Knopf, 1984.

Notehelfer, F. G. 'Japan's First Pollution Incident.' *Journal of Japanese Studies*. 1: 2 (Spring 1975): 351–83.

——'Between Tradition and Modernity: Labor and the Ashio Copper Mine.' *Monumenta Nipponica*. 39: 1 (Spring 1984): 11–24.

Ohara, Keiji, ed. Okato Tamotsu, trans. *Japanese Trade and Industry in the Meiji-Taisho era*. Tokyo: Obunsha, 1957.

Ohkawa Kazushi and Hirohisa Kohama. *Lectures on Developing Economies: Japan's Experience and Its Relevance*. Tokyo: University of Tokyo Press, 1989.

Ohkawa Kazushi, Gustav Ranis, with Larry Meissner. *Japan and the Developing Countries: A Comparative Analysis*. Oxford: Basil Blackwell, 1985.

Ohno Kennichi. *The Economic Development of Japan: The Path Traveled by Japan as a Developing Country*. Tokyo: GRIPS Development Forum, 2006.

Okita, Saburō. *The Developing Economies and Japan: Lessons in Growth*. Tokyo: University of Tokyo Press, 1980.

Ōkuma, Shigenobu, Marcus B. Huish, trans. *Fifty Years of New Japan*. New York: E.P. Dutton, 1909.

'On Abandoning the use of Leathern Boots for Japanese Soldiers.' *The Japan Times*. (2 February 1878): 95.

Ono Akira. 'Technical Progress in Silk Industry in Prewar Japan: The Types of Borrowed Technology.' *Hitotsubashi Journal of Economics*. 27 (1986): 1–10.

Ōtsuka, Keijiro, Gustav Ranis and Gary Saxonhouse. *Comparative Technology Choice in Development: The Indian and Japanese Cotton Textile Industries*. New York: St. Martin's Press, 1988.

Overman, Frederick. *A Treatise on the Metallurgy*. New York: D. Appleton, 1868.

Oxford, Wayne H. *The Speeches of Fukuzawa: A Translation and Critical Study*. Tokyo: The Hokuseido Press, 1973.

Patrick, Hugh, ed. *Japanese Industrialization and its Social Consequences*. Berkeley, CA: University of California Press, 1973.

Pauer, Erich, ed. *Silkworms, Oil, and Chips ...* Bonn: Japanologisches Seminar, 1986.

——*Papers of the History of Industry and Technology in Japan*. 3 vols. Marburg: Förderverein Marburger Japan-Reihe, 1995.

Phillips, Glyn O. *Innovation and Technology Transfer in Japan and Europe: Industry–Academic Interactions*. London: Routledge, 1989.

Phillips, J. Arthur. *Elements of Metallurgy, A Practical Treatise on the Art of Extracting Metals from the Ores*. London: Charles Griffin, 1887.

Pinch, Trevor J. and Wiebe E. Bijker. 'The Social Construction of Facts and Artifacts: Or How the Sociology of Science and the Sociology of Technology Might Benefit From Each Other.' *Social Studies of Science*. 14 (1984): 339–441.

Pittau, Joseph. 'Inoue Kowashi, 1843–1895, and the Formation of Modern Japan.' *Monumenta Nipponica*. 20: 3/4 (1965): 253–82.

——*Political Thought in Early Meiji Japan, 1868–1889*. Cambridge, MA: Harvard University Press, 1967.

Porter, Theodore. *The Rise of Statistical Thinking, 1820–1900*. Princeton, NJ: Princeton University Press, 1986.

Pownall, Charles A. W. 'The Railway System of Japan.' *Transactions of the Institution of Civil Engineers*. 124 (1896): 470–5.

'Progress–Japan and China.' *Japan Weekly Mail*. (23 September 1871): 545–6.

Pumpelly, Raphael. *Geological Researches in China, Mongolia and Japan, During the Years 1862–1865*. Philadelphia, PA: Collins, 1866.

Purcell, Hugh G., ed. *Japan Journal: The Private Notes of Gervaise Purcell, 1874–1880*. Alhambra, CA: Meiji Art/Litho, 1975.

Pye, David. *The Nature of Aesthetics of Design*. London: Herbert Press, 1978.

Pyle, Kenneth B. *The New Generation in Meiji Japan: Problems of Cultural Identity, 1885–1895*. Stanford, CA: Stanford University Press, 1969.

Ranade, Mahadeva G. *Essays on Indian Economics*. Bombay: Thacker, 1898.

Rawlinson, John L. *China's Struggle for Naval Development, 1839–1895*. Cambridge: Harvard University Press, 1967.

Rein, Johannes J. *Industries of Japan*. London: Hodder and Stoughton, 1889.

'Report on the Iron and Steel Industries of Foreign Countries: Japan.' *Journal of the Iron and Steel Institute*. (1880, part 1): 381–3.

'The Return of the Embassy.' *Japan Weekly Mail*. (23 September 1873): 594–6.

'The Return of Japanese Students From Abroad.' *Japan Weekly Mail*. (19 June 1874): 380–1.

'The Return of Japanese Students From Abroad.' *Japan Weekly Mail*. (8 July 1874): 489.

Ricoeur, Paul. *From Text into Action*. Kathleen Blamey and John B. Thompson, trans., Evanston, IL: Northwestern University Press, 1991.

Roberts, David E. 'The Development of Blast-Furnace Blowing-Engines.' *Proceedings of the Institution for Mechanical Engineering*. Nos. 3 and 4 (1906): 375–87.

Roberts-Austin, W. C. *An Introduction to the Study of Metallurgy*. London: Charles Griffin, 1910.

Robinson, Richard H. *The International Transfer of Technology: Theory, Issues, Practice*. Cambridge, MA: Ballinger Publishing Company, 1988.

Roe Smith, Merritt. *Does Technology Drive History? The Dilemma of Technological Determinism*. Cambridge: MIT Press, 1994.

Ross, Kristin. *Fast Cars, Clean Bodies: Decolonization and the Reordering of French Culture*. Cambridge, MA: MIT Press, 1995.

Rothwell, Richard P. Ed. *The Mineral Industry, its Statistics, Technology and Trade in the United States and Other Countries for the Earliest Times to the end of 1892: Statistical*

Supplement of the Engineering and Mining Journal. Vol. 1. New York: The Scientific Publishing Co., 1893.

Rule, James B. *Theory and Progress in Social Science.* Cambridge: Cambridge University Press, 1997.

Ryosuke, Ishii. *Japanese Legislation in the Meiji Era.* 1958.

Sakata, Yoshio and John Whitney Hall. 'The Motivation of Political Leadership in the Meiji Restoration.' *Journal of Asian Studies.* 16: 1 (Nov., 1956): 31–50.

Samuels, Richard J. *Rich Nation, Strong Army.* Ithaca, NY: Cornell University Press, 1994.

Saxonhouse, Gary. 'A Tale of Japanese Technological Diffusion in the Meiji Period.' *The Journal of Economic History.* 34: 1 (March 1974): 149–165.

——'Determinants of Technology Choice: The Indian and Japanese Cotton Industries.' In *Japan and the Developing Countries, A Comparative Analysis.* Ohkawa Kazushi and Gustav Ranis, eds. Oxford: Basil Blackwell, 1987.

Schatzberg, Eric 'Ideology and Technical Choice: The Decline of the Wooden Airplane in the United States, 1920–1945.' *Technology and Culture.* 35: 1 (1994): 34–69.

——*Wings of Wood, Wings of Metal: Culture and Technical Change in American Airplane Materials, 1914–1945.* Princeton, NJ: Princeton University Press, 1999.

Schubert, H. B. *History of the British Iron and Steel Industry, from c. 450 BC to AD 1775.* London: Routledge and Kegan Paul, 1957.

Sheridan, George J. Jr, *The Social and Economic Foundation of Association Among the Silk Weavers of Lyons, 1852–1870.* 2 vols. New York: Arno Press, 1981.

Shibusawa Keizo, ed. *Japanese Life and Culture in the Meiji Era.* Charles S. Terry, trans. Tokyo: Pan-Pacific Press, 1958.

Shils, Edward. 'Ideology: The Concepts and Function of Ideology.' In *International Encyclopedia of the Social Sciences.* Vol. 7, Sills, David., ed. New York: Crowell Collier and MacMillan, 1968, pp. 66–75.

'Shipping Intelligence.' *Japan Weekly Mail.* (26 March 1874): 193.

'Shipping Intelligence–Outwards.' *Japan Times.* (16 February 1878): 142.

Shiro Sugihara and Toshihiro Tanaka, eds. *Economic Thought and Modernization in Japan.* Northampton, MA: Edward Elgar, 1998.

Shively, Donald H., ed. *Tradition and Modernization in Japanese Culture.* Princeton, NJ: Princeton University Press, 1971.

Siegfried, Erich. *Das Korps Franconia in Freiberg, 1838–1910.* Leipzig: Druck von Breitkopf & Härtel, 1910.

Silberman, Bernard S. 'Bureaucratic Development and the Structure of Decision-Making in the Meiji Period: The Case of the Genro.' *Journal of Asian Studies.* 27: 1 (Nov., 1967): 81–94.

Silberman, Bernard S. and H. D. Harootunian, eds. *Modern Japanese Leadership; Transition and Change.* Tucson, AZ: University of Arizona Press, 1966.

Sims, Richard. *French Policy Towards the Bakufu and Meiji Japan, 1854–95.* Richmond: Japan Library (Curzon Press Ltd.), 1998.

Smiles, Samuel. *Self Help with Illustrations of Conduct and Perseverence.* London: John Murray, 1859 (1958 centenary edition, intro by Asa Briggs).

Smith, Thomas C. 'The Introduction of Western Industry to Japan During the Last Years of the Tokugawa Period.' *Harvard Journal of Asiatic Studies.* 11: 1/2 (June, 1948): 130–52.

——*Political Change and Industrial Development in Japan: Government Enterprise, 1868–1880.* Stanford, CA: Stanford University Press, 1955.

Sohō Tokutomi. *The Future Japan*. Vinh Sihn, trans. and ed., Matsuzawa Hiroaki and Nicholas Wickenden, co-eds. Ontario: University of Alberta Press, 1989.

Spence, Jonathan. *The Search for Modern China*. New York: Norton, 1990.

Spencer, Herbert. *The Works of Herbert Spencer, Essays: Scientific, Political, and Speculative*. Vol. 13, Onsabruck: Otto Zeller, 1966 (reprint of 1891 edition).

'Status of Charcoal Iron.' *Journal of the United States Association of Charcoal Iron Workers*. 2: 4 (1881): 228.

Stobaugh, Robert and Louis T. Wells, Jr., eds. *Technology Crossing Borders: The Choice, Transfer, and Management of International Technology Flows*. Boston, MA: Harvard Business School Press, 1984.

'The Study of English by Japanese.' *Japan Weekly Mail*. (25 November 1871): 656–8.

'Successful Iron Works.' *Journal of the United States Charcoal Iron Workers*. 3: 4 (1882): 234–6.

Sugihara, Shiro and Toshiro Tanaka, eds. *Economic Thought and Modernization in Japan*. Northampton, MA: Edward Elgar, 1998.

Sugiyama Chuhei. *Origins of Economic Thought in Modern Japan*. London: Routledge, 1994.

Sugiyama Chuhei and Hiroshi Mizuta. *Enlightenment and Beyond: Political Economy Comes to Japan*. Tokyo: University of Tokyo Press, 1988.

Sugiyama Shinya. *Japan's Industrialization in the World Economy 1859–1899: Export Trade and Overseas Competition*. London: Anthlone Press, 1988.

Suzuki Jun. 'The Humble Origins of Modern Japan's Machine Industry.' In Masayuki Tanimoto, ed. *The Role of Tradition in Japan's Industrialization: Another Path to Industrialization*. Oxford: Oxford University Press, 2006.

Swank, James M. *Statistics of the American and Foreign Iron Trades*. Philadelphia, PA: The American Iron and Steel Institute, 1878.

——*Statistics of the American and Foreign Iron Trades for 1899*. Philadelphia, PA: The American Iron and Steel Institute, 1900.

Takahashi Kamekichi. *The Rise and Development of Japan's Modern Economy: The Basis for 'Miraculous' Growth*. John Lynch, trans. Tokyo: The Jiji Tsushinsha, 1969.

Takeuchi, Johzen. *The Role of Labour-Intensive Sectors in Japanese Industrialization*. Tokyo: United Nations University Press, 1991.

Takeuchi, Tsuneyoshi. *The Formation of the Japanese Bicycle Industry: a Preliminary Analysis of the Infrastructure of the Japanese Machine Industry*. Tokyo: United Nations University Press, 1981.

Tanimoto Masayuki, ed. *The Role of Tradition in Japan's Industrialization: Another Path to Industrialization*. Oxford: Oxford University Press, 2006.

Tanimura Hiromu. 'Development of the Japanese Sword.' *Journal of Metals*. 32: 2 (February, 1980): 63–73.

Teeters, Barbara J. 'Kuga's Commentaries on the Constitution of the Empire of Japan.' *Journal of Asian Studies*. 28: 2 (February 1969): 321–37.

Tomoko Hashino and Osamu Saito, 'Tradition and Interaction: Research Trends in Modern Japanese Industrial History.' *Australian Economic History Review*. 44: 3 (November 2004) 241–258.

Tsuchida, Kyoson. *Contemporary Thought of Japan and China*. New York: A.A. Knopf, 1927.

Tsunehiko Yui and Keiichiro Nakagawa, eds. *Japanese Management in Historical Perspective: Proceedings of the Fuji Conference*. Tokyo: University of Tokyo Press, 1989.

Tsunoda, Ryusaku, Wm. Theodore De Bary, and Donald Keene. *Sources of Japanese Tradition*. Vol. 2. New York: Columbia University Press, 1964.

Tsunoyama, Sakae. *A Concise Economic History of Modern Japan*. Bombay: Vora & Co., Publishers Private Ltd., 1965.

Tsurumi, Yoshi. *Japanese Business: a Research Guide with Annotated Bibliography*. New York: Praeger, 1978.

Umegaki Michio. *After the Restoration*. New York: New York University Press, 1988.

Unotoro Shingo. 'The Progress and Present State of Pig Iron Manufacture in Japan.' *Proceedings of the World Engineering Conference*. 33: 1 (1931): 267–89.

'Unsound Progress.' *The Japan Weekly Mail*. (30 December 1871): 719–21.

Uyehara, Shigeru. *The Industry and Trade of Japan*. London: P. S. King, 1936.

Vincenti, Walter G. *What Engineers Know and How They Know It: Analytical Studies from Aviation History*. Baltimore, MD: The Johns Hopkins University Press, 1990.

von Clausewitz, Carl. *On War*. Edited and translated by Michael Howard and Peter Paret. Princeton, NJ: Princeton University Press, 1984.

von Siebold, Alexander. *Japan's Accession to the Comity of Nations*. London: Kegan Paul, Trench, Trübner & Co. Ltd., 1901.

Wakabayashi, Robert Tadashi. *Anti-Foreignism and Western Learning in Early-Modern Japan: The New Thesis of 1825*. Cambridge, MA: Council on East Asian Studies, Harvard University, 1991.

——ed. *Modern Japanese Thought*. Cambridge, MA: Cambridge University Press, 1998.

Ward Robert, ed. *Political Development in Modern Japan*. Princeton, NJ: Princeton University Press, 1968.

Weber, Max. *Economy and Society: An Outline of Interpretive Sociology*. Guenther Roth and Claus Wittich, eds. New York: Bedminster Press, 1968.

Wedderburn, David. 'Modern Japan.' *The Fortnightly Review*. Vol. 23, new series (1 January–1 June, 1878): 417–542.

Westney, Eleanor D. *Imitation and Innovation: The Transfer of Western Organizational Patterns to Meiji Japan*. Cambridge, MA: Harvard University Press, 1987.

Willoughby, Kelvin W. *Technology Choice: A Critique of the Appropriate Technology Movement*. Boulder, CO: Westview Press, 1990.

Wittner, David G. *Technology Transfer in the Meiji Era: The Case of Kamaishi Iron Works, 1874–1890*. Unpublished Master's Thesis, Columbus, OH: The Ohio State University, 1995.

Wood, Nicholas. 'On the Conveyance of Coals in Underground Coal Mines.' *Transactions of the North England Institute of Mining Engineers*. 3 (1854 &1855): 239–80.

Wray, William D., ed. *Managing Industrial Enterprise: Cases from Japan's Prewar Experience*. Cambridge, MA: Harvard University Press, 1989.

Yamaguchi, Kazuo. 'The Leaders of Industrial and Economic Development in Modern Japan.' *Cahiers D'Histoire Mondiale*. 9: 2 (1965): 233–53.

Yamamoto, Shichihei. Lynne E. Riggs and Takeuchi Manabu, trans. *The Spirit of Japanese Capitalism and Selected Essays*. Lanham, MD: Madison Books, 1992.

Yamamura, Kozo. 'Success Illgotten? The Role of Meiji Militarism in Japan's Technological Progress.' *Journal of Economic History*. 37: 1 (March 1977): 113–35.

Yanagida Kunio, ed. *Japanese Manners and Customs of the Meiji Era*. Charles S. Terry, trans. Tokyo: Pan-Pacific Press, 1957.

Yokoyama, Toshio. *Japan in the Victorian Mind: A Study of Stereotyped Images of a Nation, 1850–80*. London: Macmillian, 1987.

Yonekura, Seiichiro. *The Japanese Iron and Steel Industry: Continuity and Discontinuity, 1850 to 1970*. Ph. D. dissertation, Harvard, 1990.

——*The Japanese Iron and Steel Industry, 1850–1990: Continuity and Discontinuity.* New York: St. Martin's Press, 1994.

Yoshihara Kunio. *Japanese Economic Development.* Third Edition. Oxford: Oxford University Press, 1994.

Yukihiko Kiyokawa. 'Transplantation of the European Factory System and Adaptations in Japan: The Experience of the Tomioka Model Filature.' *Hitotsubashi Journal of Economics.* 28 (1987): 29.

Index

Abe Kiyoshi 118
Adams, F. O. 51
agriculture 27, 115–16
Aiichi Spinning Mill 36, 37, 121*t*
Akabane Machine Works 36, 105, 108
Akasaka filature 53, 54–5, 59, 104–5, 129
Alcock, Sir Rutherford 104
Ani Copper Mine 34, 121*t*
architecture 101, 162 n23
arms-for-silk deals 46–7
Asano Sōichiro 118–19, 120
Ashio copper mines 34, 119
Ayrton, William Edward 29

Bakumatsu era 28
Bianchi, Louis 29, 78, 79–80, 81, 82,
    83–4, 86, 87, 91, 130
British influences: cotton industry 36–8;
    education 29–30, 31; iron industry 13,
    77–8, 82, 84, 87, 89–91, 92; mining
    34–5, 89–90; railroads 30–1; silk
    industry 11, 46, 47, 89; telegraph
    system 32
Brunat, Paul 44, 48, 49–50, 55, 59, 64, 69,
    70, 116, 146 n33
Brunton, Richard Henry 30, 93, 95, 150
    n4, 164 n52
bunmei kaika see civilization and
    enlightenment
Bureau of Taxation 54
Bury, J. B. 9

Casley, William H. B. 87
charcoal 85, 87, 88
China 11, 111–12, 123
Chōshū Five 29
civilization: definitions 9, 122
civilization and enlightenment (*bunmei
    kaika*) 8–9, 99–124, 126–7; 1868–77:

Meiji techno-diplomacy 101–6;
    1878–1890s: selective techno-political
    restructuring 106–9, 128; 1890s:
    rejection of *bunmei kaika* 109–24;
    civilizational hierarchy 101, 104,
    125–6; criticisms 13–14, 15, 17–18,
    92–3, 95–6, 122; cultural materiality
    99–100, 101; ideology 15, 16, 17, 100,
    101; Matsukata Deflation 14, 69, 107;
    physical manifestations 15–16, 100–4,
    105; stages of civilization 100, 122;
    terminology 8–9, 99 *see also* progress
    ideology
Clancey, Gregory K. 55, 162 n23
coal 34–5, 48, 72, 80, 83, 85, 88–9
Coignet, Francisque 32–3, 34
coke 88, 89
College of Engineering (*Kōbu daigakkō*)
    29–30, 31, 34, 35
Conder, Josiah 29, 128, 162 n23
conservatism 112, 167 n99
copper 34, 72, 73, 92
cotton spinning industry 30, 35, 36–8, 39,
    108
Coye, Ainé 56, 59
cultural materiality 4–5, 12, 95, 96,
    99–100, 101, 123, 125–6, 127–8

Date Munenari 11, 51
diet 103
Dohmen, Martin 149–50 n110
Dubousquet, Albert Charles 44
Dunham, Wheeler 87
Dyer, Henry 29, 30

economy 107, 165 n65; agriculture
    115–16; disposal of state-owned
    enterprises 107–9, 115, 121*t*, 128–9,
    165 n76; Kamaishi Ironworks 91–2,

93, 94, 97, 108, 115, 119, 120; military
    31, 112, 116–17; silk reeling industry
    38, 49–50, 53, 69
educational system 28, 29–30, 31, 42
Emperor Meiji 102–3
Endō Kinsuke 29
Engineer Training College 31, 93
engineering 29–30, 42
Engineering Bureau (*Kōgakuryo*) 29
Ericson, Steven 94, 143 n53
Evans & Haskins 84–5
exhibitions, national and international 3,
    15, 16, 47, 55, 64–5, 69, 70*t*, 104

fashion 96–7, 102–3, 170 n154
Federico, Giovanni 149 n100
Forbes, David 84–6
France: silk industry 11, 12, 44, 46, 64–6;
    technological advisers 28, 44
Francks, Penelope 49
Friedel, Robert 5
Fukagawa Brick Factory 36, 108, 120,
    121*t*
Fukugawa Cement Factory 118
Fukuzawa Yukichi 8, 9, 99, 100, 101, 123,
    128, 161 n8, 167 n104
Furukawa Ichibē 34, 119

garabō (rattling spindle) 39
Gaun Tokimune 39
Geisenheimer, Freiderich 44, 47, 67
Glover, Thomas Blake 34–5
Godai Tomoatsu 36, 119
Godfrey, J. G. H. 32, 77–8, 79, 80, 90, 91,
    93, 97
gold 33, 72, 83, 92
Gotō Shōjirō 3435
government enterprises: disposal of
    107–9, 115, 121*t*, 128–9, 165 n76
Gower, Erasmus 33

handicraft industries 27
Harada Katsumasa 162 n26
Harada Noriko 138 n35
Hasegawa Yoshimichi 87
Hattori Shun'ichi 38
Hayami Kenzō 50–2, 53, 54, 64, 67, 68, 69
Hayes, Carlton J. H. 112
Hècht, Lilienthal and Co. 44, 45*f*, 46, 47
Hérand, Flury 47
Hirata Yonekichi 39
Hokkaidō land deal 119
Holme, Edward 37
Hosokawa Junjirō 102

ideology of technological progress
    see progress ideology
Ikuno Silver Mine 32, 34, 121*t*
Imo Mutiny (1882) 112
Imperial Charter Oath 15
imperialism: Japanese imperialism 106,
    112, 129, 172 n11; techno-imperialism
    127, 172 n10; Western imperialism 17,
    107, 123
industrial modernization 10–11, 26–42;
    cotton spinning 30, 35, 36–8, 39, 108;
    criticisms 31, 32, 95; disposal of
    government enterprises 107–9, 115,
    121*t*, 128–9, 165 n76; education 28,
    29, 42; engineering 29–30, 42; foreign
    advisers 28–9, 30, 31, 32, 34–5;
    foreign knowledge 26–7, 28, 42;
    fukoku kyōhei 27; government
    ministries 27–8; military program 28;
    mining 32–5; model factories 35–6,
    105, 107; railways 30–1; *shokusan
    kōgyō* 27, 28, 42; telegraph system 32;
    traditional industries and attitudes 38–9
    see also iron industry; silk reeling
    industry: mechanization
Inokuchi Ariya 30
Inoue Kaoru 29, 68, 72, 108, 119, 120, 140
    n61, 148 n77
Inoue Masaru 29
Iron and Steel Institute 85
iron industry: armaments 111, 124;
    bridges 72–3; cast iron 55, 74, 101–2;
    imports 117–18, 117*t*; military
    importance of 105, 109, 110, 111,
    116–17; modernization 3–4, 6, 7, 90–1,
    92; national importance of 110–12,
    113, 115; pig iron 25, 26, 74; progress
    ideology 7, 13, 72–3, 89, 95–6, 105–6,
    109 *see also* iron smelting and mining:
    traditional technologies; Kamaishi
    Ironworks
iron smelting and mining: traditional
    technologies 2, 24–6; bellows 25–6,
    83; extraction 24; iron ore 24, 26; iron
    sand 24–5; *kanna nagashi* (water
    method) 24; *kera* (steel and iron) 25;
    knife iron 25; *nodatara* (furnace) 25;
    pig iron 25, 26; smelting 25–6; *tama
    hagane* (steel) 25; *tatara* furnaces
    25–6
Italy 12, 51, 65, 66–7
Itō Hirobumi 15, 17, 27–8, 29, 31, 42, 44,
    67, 78, 86, 90, 91, 92, 93, 95, 102, 109,
    112, 140 n60

Itō Miyoji 119
Ito Moemon 61*f*, 63*f*
Itō Yajirō 94, 95, 118
Iwakura Mission 2, 15, 35, 66, 77, 104,
    122
Iwakura Shoko Tomomi 112
Iwasaki Yatarō 34

Jardine, Matheson & Co. 35
Jiji shinbun 118
Jiro Kumagai 167 n104
Jukki Spinning Mills 36, 37

Kagoshima Cotton Spinning Mill 35, 36,
    37
Kamaishi area furnaces 75, 75*t*, 76 *map*,
    77–8, 81, 91
Kamaishi Ironworks 73–98, 76 *map*;
    1854–74: early years 74–5;
    1874–83: public works era 74, 77–89,
    118; 1883–1924: disposal 17–18, 74,
    94–8, 113, 115, 118–19, 120, 121*t*,
    128–9; construction 83–4, 86–7;
    design 78–9, 81–3, 84, 85, 93;
    economics 91–2, 93, 94, 97, 108,
    115, 119, 120; equipment 84,
    85–6; fuel 85, 87, 88–9; location
    proposals 80–1, 86–7, 130–1;
    negative assessment 90–2, 93,
    95, 96, 97; production 87–8; and
    progress ideology 13–14, 16–17,
    92–3, 97–8, 106; publicity 79, 93, 94,
    110; purpose 3, 13, 92; railroad 78–9,
    84, 86, 87, 88, 92, 94; Western
    technology 14, 82, 84, 85; workforce
    84
Katakura & Co. 113*f*, 114*f*
Kawamura Yoshia 128
Kawase Hideharu 122–3
Kido Takayoshi 68
Kikuchi Kyōzō 30, 38
Kōbushō enkaku hōkoku 53
Kōgyō iken ('Views on the Promotion of
    Industry') 116–17
Koma Rinosuke 79, 86, 87, 91
Komura Koroku 95
Konjaku Monogatari 24
Korea 17, 106, 111–12, 123
Kuga Katsunan 167 n103
Kume Kunitake 101, 104
Kuroda Kiyotaka 119
Kuwabara Masa 91

Li Hongzhang 111

Lilienthal, Sigismond 46
London Crystal Palace Exhibition (1851)
    104
Lyman, Benjamin Smith 32, 90–1, 95, 97

McCallion, Stephen 52, 59, 137 n9, 146
    n23
machine tool industry 28
Maebashi filature 51–2, 59, 64, 67, 69, 70
Maeda Masana 116
Maejima Hisoka 102
Mann, Matthew D. 87
Maquet, Jacques 18, 139 n53
Maruyama Yasaburō 42
material culture 6, 7–8
Matsukata Deflation 14, 69, 107
Matsukata Masayoshi 14, 18, 54, 69, 97,
    107, 115–16, 118, 120, 122, 123
Matsumura Harusuke 99
mechanization: terminology 10
Meiji Restoration 47
Mezger, Adolf 34
military: armaments industry 111, 124;
    expenditure 31, 112, 116–17;
    importance of iron 105, 109, 110, 111,
    116–17; industrial modernization 28;
    uniform 102–3
Milne, John 29
mining 32–5, 72; British interests 34–5,
    89–90; coal 34–5, 80, 83, 88; copper
    34; criticisms 93; extraction 24, 32–4;
    gold 33; iron 24–6; laws 90, 92;
    managerial system 33, 34; ownership
    3–4, 32, 33, 35; refining 33; silver
    32–3; technical training 34
Mining Department 79, 80, 83, 84, 90
Ministry of Agriculture and Commerce
    (*Nōshōmushō*) 27, 69, 116
Ministry of Civil Affairs (*Minbushō*) 27,
    35, 44, 47–8, 53, 64, 68, 70
Ministry of Education (*Monbushō*) 29
Ministry of Finance (*Ōkurashō*) 27, 35, 37,
    44, 47–8, 53, 64, 94, 97, 111, 116
Ministry of Foreign Affairs (*Gaimushō*) 51
Ministry of Home Affairs (*Naimushō*) 27,
    28, 35, 53, 54, 108, 122
Ministry of Public Works (*Kōbushō*) 27,
    28, 29, 30, 32, 34, 35, 56, 59, 94, 104,
    105, 108, 109, 111; Mining Department
    79, 80, 83, 84, 90; Office for the
    Promotion of Industry 53, 54; Railway
    Department 80, 84, 94
Ministry of the Army 28, 109, 111
Ministry of the Navy 28, 109, 111

Mito furnace 74, 75, 81
model factories 35–6, 105, 107
modern: definitions 9, 164 n58
modernization 1–2, 7–8; terminology 10,
   18, 164 n58; Western influence 15–16
Montblanc, Comte de 47
Mori Jusuke 87
Motoda Eifu 17, 164 n62
Motoyama Yukihiko 55
Mueller, Casper 51, 53, 64, 69
mutual-progress exhibitions (*kyōshinkai*)
   16

Nabeshima Naomasa 4, 34
Nagai Yasuoki 56, 59
Nagasaki Shipyard 13, 121*t*
Nakakosaka Iron Mine 120, 121*t*
Nakamura Masanao 99
Nakamura Michita 48
national identity 110–11, 112, 122–4
national security 109, 110, 111, 112–13,
   113*f*, 114*f*, 116
nationalism 17, 112
Naumann, Edmund 32
Netto, Curt 32, 33, 34, 91–2, 97
Nihon Ryoiki 24
Niiro Gyōbu 36
Nippon Railway 31
Nishimura Shigeki 15, 109
Noro Kageyoshi 95

Obana Fuyukichi 110–11
Odaka Atsutada 48, 55, 64, 69, 70, 146–7
   n33
Office for the Promotion of Industry 53, 54
Ohara Zengoro 74
Ōhashi ironworks 75, 75*t*, 79, 81, 88, 91
Oki Takato 48
Ōkubo Toshimichi 15–16, 28, 35–6, 42,
   101, 171 n2
Ōkuma Shigenobu 31, 36, 44, 102, 104,
   107–8, 119, 172 n9
Osaka Cotton Spinning Company 37–8
Ōshima Takatō 74–5, 77, 78, 79, 80–3, 84,
   86, 87, 95, 97, 130, 131

Paris International Exhibition (1867) 15,
   47, 64–5, 104
Parkes, Sir Harry 11, 30, 46, 47, 90, 97,
   102
Percy, John 84
Perry, Matthew C. 9, 30, 32
Platt Brothers 36, 37
Plunkett, F. W. 90

progress: definitions 9
progress ideology 10, 127–8; definition
   6–7; in the iron industry 7, 13, 72–3,
   89, 95–6, 105–6, 109; Kamaishi
   Ironworks 13–14, 16–17, 92–3, 97–8,
   106; in the silk reeling industry 7, 12,
   49, 52–3, 103; Tomioka Silk Filature
   12, 16, 43, 52–3, 55, 67–9, 103–4
Pumpelly, Raphael 33
Purcell, Gervaise 84, 86, 87
Putiatin, E. V. 30
Pyle, Kenneth B. 109–10, 123, 164 n62,
   167 n99

Railway Department 80, 84, 94
railways 30–1, 94, 100, 102, 162–3 n28; at
   Kamaishi 78–9, 84, 86, 87, 88, 92, 94
Rein, Johannes J. 95
Ricoeur, Paul 137 n16
Rikuchū Heigori iron mines 73, 77, 81, 91
Roches, Leon 46–7, 65, 104
Rokkōsha filature 52
Rokumeikan 128

Sado Gold Mine 34, 83, 88, 121*t*
Saga Coal Mine 4, 34–5
Saigo Takamori 71, 162–3 n28
Saito Toichi 39
Saitō Tsunezō 38
Sakai Cotton Spinning Mill 36, 37
Sano Tsunetami 3, 16, 55, 69, 70*t*, 73, 103,
   104, 106
Sapporo Machine Works 105
Satsuma–Chōshū alliance 119, 120, 123
Schatzberg, Eric 6, 161 n10
selective techno-political restructuring
   106–9, 128
Senjū Woolen Mill 36, 108
sericulture and silk industry 2; British
   influence 11, 46, 47, 89; China 11;
   Europe 11; exports 11, 46, 46*t*, 103;
   France 11, 12, 44, 46, 64–6; Italy 51,
   65, 66; looms 39; pébrine virus 11, 51,
   66; Yokohama 11 *see also* silk reeling;
   traditional technologies; silk reeling
   industry: mechanization
Shibusawa Eiichi 15, 37, 38, 44, 48–9, 54,
   64–5, 66, 69, 70, 96, 104, 118–19, 120
Shiga Shigetaka 164 n62
Shimazu Nariakira 32, 36, 120
Shinagawa Glass Factory 36, 121*t*
Shinmachi Waste Silk Reeling facility
   108, 128
shipbuilding 13, 28, 92, 121*t*

Shively, Donald H. 109
silk reeling: traditional technologies 2, 19–23; *doguri* (reeling device) 19; *noshi* (coarse thread) 19, 21; re-reeling 23, 23*f*; *tebiki* (reeling device) 19; *zaguri* (reeling frame) 19–20, 20*f*, 21*f*, 22*f*, 23
silk reeling industry: mechanization 2–3, 6, 44–71; Akasaka filature 53, 54–5, 59, 104–5, 129; British influence 11, 46, 47, 89; *chambon* (*toyomori*) *croisure* 59, 60*f*, 61*f*; choice of technique 47–50, 59, 64–7; construction materials 55, 58*f*, 59; criticisms 49–50; economic perspectives 38, 49–50, 53, 69; filatures: distribution and productivity (1895) 132–5; initial contact 7, 44–7; machinery 55–8, 56*f*, 57*f*, 59; Maebashi filature 51–2, 59, 64, 67, 69, 70; materiality of 'civilization' 50–9, 96, 102, 104–5; progress ideology 7, 12, 49, 52–3, 103; reelers' manuals 56, 56*f*, 59; Rokkosha filature 52; steam 42; *tavelle (kenneru) croisure* devices 39, 40*f*, 41*f*, 59, 62*f*, 63*f*; terminology 10; 'Tomioka-style' 52; validation and idealism 67–9 *see also* Tomioka Silk Filature
silver 32–3, 72, 83, 84, 91, 92, 165 n65
Sims, Richard 46, 138 n39
Smith, Thomas C. 165 n76
Société Générale Française d'Exportation et d'Importation 46–7
Soeda Juichi 111
steel imports 117–18, 117*t*
Sugiura Yuzuru 15, 48, 49, 64–5, 66, 104
Switzerland 65
symbolism 4, 43, 71, 128

Tachi Saburō 60*f*, 62*f*
Takamine Jokichi 30
Takashima Coal Mines 34–5, 121*t*
Takatsuji Narazo 38
Tamanoe Seiri 48
Tanaka Chōbei 94, 95, 120, 122
Tatsuno Kingo 30
techno-diplomacy 101–6, 121–6, 128
techno-imperialism 127, 172 n10

technological artifacts: and cultural values 5–6, 8, 126–7; design and selection 5; as texts 5
technological determinism 6, 129
technological progress 6 *see also* progress ideology
technology transfer 2, 28–9
telegraph system 32, 102
terminology 6–7, 8–10, 18, 164 n58
Tokugawa Akitake 64, 66
Tokugawa Nariakira 75
Tokutomi Sohō 128, 164 n62, 172 n9
Tokyo Imperial University 29
Tokyo nichinichi shinbun 100, 103
Tomioka Silk Filature: disposal 108, 121*t*, 128; French influence on 11, 12, 44; influence of 5, 52–3, 69; international recognition 16, 55, 103–4; progress ideology 12, 16, 43, 52–3, 55, 67–9, 103–4; purpose 3, 11–12, 43–4, 47–8, 59, 64, 67–8, 71
Toyoda Sakichi 39
traditional techniques 9–10, 38–9
transportation 30–1, 65, 100 *see also* railways
Tsuda Tsukane 37

Umegaki Michio 67

Victor Emmanuell II 66
Vienna International Exhibition (1873) 3, 55, 69, 70*t*, 104
von Siebold, Alexander 46, 47, 64, 161 n9

Wilson, George 84
Winchester, Charles 46
wood 7, 56, 58, 106, 127

Yamada Jun'an 87
Yamagata Aritomo 112
Yamanobe Takeo 37–8
Yamao Yōzō 3, 29, 42, 78–9, 81, 82, 85, 87, 91, 92, 93, 97, 109
Yokohama 11, 50–1, 150 n110
Yokohama Chamber of Commerce 51
Yokoyama Kyutarō 120
Yoshii Tomozane 48
Yoshii Tōru 77, 82–3

Lightning Source UK Ltd.
Milton Keynes UK
29 November 2009

146841UK00001B/95/P

9 780415 560610